Writing Acupuncture
Case Reports

Writing Acupuncture Case Reports

THEORY AND PRACTICE

Edward Chiu, LAc, DAOM

Library of Congress Cataloging-in-Publication Data

Chiu, Edward

Writing acupuncture case reports: theory and practice / Edward Chiu, LAc, DAOM

Includes bibliographical references.

ISBN 978-1-7359583-0-9 (paperback: alk paper).

1. Acupuncture 2. Integrative medicine 3. Technical writing. 1. Title

Library of Congress Control Number: 2020921934

Book design by Cecile Kaufman

Cover Artwork by Sunjae Lee

Printed in the United States of America

Table of Contents

Introduction

From all appearances, acupuncture is entering a golden age. Modern research has elucidated a variety of physiological mechanisms that validate the effects that acupuncturists have observed for centuries, and hundreds of clinical trials have proven efficacy of acupuncture treatment for a variety of conditions. More and more patients are pursuing and demanding acupuncture as a complement to their biomedical care, and hospitals and insurance companies are responding by making acupuncture a part of their treatment offerings. Awareness and availability of acupuncture are both rising, and it would seem that it is on the verge of widespread acceptance.

Nevertheless, within wide swaths of the biomedical community and the population in general, traditional theories of acupuncture are still criticized as pseudoscience. A growing number of allopathic practitioners are performing acupuncture procedures with minimal training and without exposure to the traditional knowledge base and its perspective. At the same time that traditional acupuncture theory is discredited as unscientific, techniques traditionally practiced by acupuncturists are being rebranded and adopted as treatment modalities by physical therapists, massage therapists, and chiropractors. Cupping is relabeled as myofascial decompression and shiny new instruments and new trademarks retool the *guasha* scraping technique. Promoters of dry needling even disparage acupuncture in print despite the fact that dry needling is one of the many techniques that traditional acupuncturists actually perform. It seems that while scientific knowledge and public consciousness about acupuncture is growing, interest in its traditional perspective is losing ground.

In biomedicine good health is a regarded as a state of balance known as homeostasis, where symptoms can ensue with too much of a lean in one direction or another. Traditional acupuncture also strives for a state of balance — and in similar fashion, disease occurs when this balance is disturbed. Some of the most basic tenets of acupuncture describe the balance in binary terms. For example, if a patient exhibits too many Heat signs, treatment is designed to give a gentle push in the cooling direction. But some of the ideas motivating acupuncture are considerably more complex; for example, five-phase theory explains how an imbalance in any one organ can affect the function of other organs. A constellation of seemingly unrelated signs and symptoms when assessed by a qualified acupuncturist can collectively indicate a pattern of imbalance to be rectified by treating a specific combination of points.

Through these and other traditional principles, acupuncture offers an alternative pathway for viewing health and its balance in the human body. This paradigm of traditional acupuncture is based on thousands of years of observation. Each patient presents an opportunity to test a prediction regarding health and balance. The knowledge gleaned from observations over the years combined with careful consideration about the outcomes of treatments has led to the many theories layered into the tradition of acupuncture.

Many acupuncturists in the modern era pursue their training because it offers a holistic route to solving health problems that personally resonates with them. Some choose to study acupuncture because of its seamless integration of mind and body. Some choose to enter the field because they appreciate how acupuncture can be flexibly adapted to treat patients as unique individuals. Some are fascinated by the methods acupuncturists use to detect imbalance by locating tense points or nodules on the acupuncture channels or by analyzing subtle changes in the strength or quality of a patient's pulse. The study of acupuncture is challenging and complex. Rather than dividing the body and its conditions into a variety of specialties, acupuncture requires its practitioners to pay attention to the interactions between the parts, and how each functionally contributes to the health of the whole person. Acupuncturists explore a constellation of seemingly unrelated symptoms to find a pattern, and by using the idea of balance as a central principle, treatments can be elegant, with even startling results. As such, the utility of acupuncture has enabled its traditional methodology to survive the many political, cultural, and scientific criticisms that it has endured for thousands of years.

In the last fifty years, researchers have tried to discover the reasons behind acupuncture's effects, and to evaluate the efficacy of acupuncture treatments. Two broad categories of acupuncture research exist. The first — mechanistic research — involves identifying and measuring specific effects when a needle is placed in the body; for example, if the blood chemistry changes or if certain parts of the brain become more active. When such an effect is consistently measured, the procedure has verifiable biological effects. The second — clinical research — seeks to measure outcomes for a trial group of patients treated with an acupuncture point protocol designed for a specific condition. When a benefit is consistently measured, the procedure of acupuncture is validated. The scientific study of acupuncture has done much to change its image from quackery to accepted and respected complementary therapy. But with that validation something has been lost: the version of acupuncture validated by mechanistic research has eliminated its significant traditional roots. In this rendering, acupuncture is redefined as merely a set of physiological reactions rather than a unique system of diagnosis and treatment based on thousands of years of observation and experience. As a result, while acupuncture as a procedure is validated, the paradigm based on the traditional theories of acupuncture appears less relevant. The alternative lens of traditional observation, diagnosis, and treatment is disparaged, discounted, and misunderstood.

For the past hundred years, the scientific method has been the primary tool which we have culturally settled upon for the last hundred years to determine the validity of a hypothesis. But because the western scientific method is not a suitable tool to prove the validity of an acupuncture *theory*, much less a collection of interconnected theories, the paradigm of acupuncture is reduced to a procedure without its context — a tree without its roots. If the history and experience that acupuncture is built upon is not valued, it is only natural that standardized medical acupuncture procedures would be increasingly performed by allopathic practitioners with minimal training. Without the wisdom and experience of the past, medical acupuncturists can only recommend treatment tested by the large-scale clinical trial, which uses the same protocol for all patients with the same condition without considering whether or not the treatment actually fits the individual patient's presentation.

For decades, the large-scale clinical trial has been considered to be the strongest type of medical evidence, while the individualized case report has been relegated towards the bottom of the evidence pyramid. As such, case reports generally do not have much influence on the practice of biomedicine, aside from a couple of categories. The first of these are index cases — the first reports of a new disease entity. These can be quite significant as they alert the medical field to a new target for research and study. Second, cases can also describe adverse events to alert readers to avoid particular interventions in specific situations for fear of harming patients. Aside from these, case reports can be used to propose a hypothesis, but very few of those make their way through the various levels of investigation to clinical trial.

Acupuncture case reports often do not fit the above criteria for a variety of reasons. They are unlikely to report the discovery of a new disease entity, partly because identification of a new disease is often dependent on diagnostic testing that is not within the scope of practice of acupuncturists. Acupuncture adverse events are similarly rare because the practice is generally considered safe. And in acupuncture case reports, the types of conclusions related to hypothesis building are often quite general, making them unlikely candidates for motivating clinical trials. As a result, case reports in acupuncture have over time been pushed aside in favor of the powerful allure of scientific progress and the clinical trial. From a biomedical perspective, they seem highly unlikely to offer anything of value.

Viewed through the lens of biomedicine, it is no wonder that writing case reports seems like a mere exercise with little to no benefit to the field. But if we look at the history of acupuncture, learning from cases was one of its cornerstones of education as well as an important component of the acupuncture literature. Acupuncture cases were written to demonstrate how theory could be used to derive effective treatment and propose new currents of thought. They were used to highlight the finer points of diagnosis and explain how to manage difficult complex cases. Reading a case report also teaches practitioners how to perform acupuncture in an individualized, patient-centered way, which is a hallmark of the traditional paradigm. Also, compiling case reports and viewing them as a collective can support the formation of generalizable knowledge rooted in the traditional theoretical understanding of acupuncture. In short, learning from cases is an essential way of learning how to practice traditional acupuncture effectively.

In addition to the overlooked benefits that case reporting can have towards clinical practice, writing about cases from a traditional viewpoint may also help us to establish the value of the Chinese medicine paradigm. In the last hundred years, evaluation of the usefulness of a specific intervention has shifted from the individual patient to larger populations, with efficacy gauged by the standards of the scientific method. Current testing methodology is not able to gauge the usefulness of traditional acupuncture theories because applying the logic of these theories is a complex process which recognizes that each individual is different. Nevertheless, even if a theory by nature cannot be validated by the standards of western medicine, it may still have explanatory and predictive value that can be demonstrated in a case report. We ought not discard a system with potential value simply because it is not testable by the methods preferred by western science.

Throughout the lengthy history of acupuncture, the primary method of validation was the observation of cases. The origin of acupuncture theory was brought about by astute practitioners making inferences about health and disease based on the collective observations of millions of patient interventions. Incorrect inferences were discarded, inconsistent ones reexamined and revised, and the ones that passed the test of time have survived. While many of these inferences were handed down through generations of practitioners via the oral tradition, the details and thought processes of a good number of cases were recorded in writing. Over time these have come to comprise the case literature — an essential learning resource for traditionally trained acupuncturists to learn the art. Observation of experienced practitioners is certainly an important way to learn a practical skill, yet much can be gained through reading about the logic of current and past experts and by examining how they apply theory to the diagnosis and treatment design of individual cases.

Considering the patient as an individual is an important piece of the original approach of evidence-based medicine, which views patient care as the combination of the best evidence available with the experience to know how to make it useful to the individual patient. Despite the reputation of large-scale clinical trials as the strongest evidence for patient care, some patients may benefit most from evidence derived from a case report. Acupuncturists are not simply concerned with 'which points work best for cough,' because the best points will likely vary from person to person depending on their presentation. The clues to understanding which points are best for the individual patient are only discoverable through a model based on experience and observation.

Before the era of search engines and medical databases, the ability to sift through the mountains of case reports to find the proverbial 'needle in a haystack' that can unlock a puzzling case would be nearly impossible. Now more than ever, the idea of finding the right case report at the right moment is much more feasible. However, to fully extract the value from such a case report, it is important not to apply its findings from a purely procedural standpoint as medical acupuncturists do, but to understand how to adapt its information by considering the knowledge of traditional acupuncture theory — something that eludes allopathic practitioners with minimal training.

The paradigm of traditional acupuncture is a guide for making effective clinical choices. While its theoretical framework, based on individual building blocks of evidence in the form of cases, cannot be proven correct or incorrect, the same could be said of Darwin's theory of evolution. Evolutionary theory cannot be proven by scientific experiments, but as with acupuncture theories, it is supported by thousands of demonstrative cases. If evolutionary theory is accepted as a way to explain the changes in our natural world, perhaps it is time to treat the collective acupuncture case report literature in a similar fashion. Indeed, demonstrating how traditional acupuncture theory can be used as a basis for deriving effective treatments may reveal the utility of considering other viewpoints and the possible acceptance of previously unfamiliar ideas.

Some might object to the use of case reports on more practical grounds. Given the diversity of schools of thought and the many strands of acupuncture theory as well as the complexity of the human organism, they might argue that any feasible methodology to support traditional practice will require a lot of cases. But the recent neglect of case reports as a viable branch of research and scholarship is no argument against them; indeed, if anything it suggests that if the integrity and strengths of traditional acupuncture are not only to be preserved but continuously developed, writing case reports must be revived. Case report literature is the evidentiary backbone of the traditional practice of acupuncture. Instead of abandoning case report writing before practitioners even begin their careers, the field of acupuncture needs to create a culture of rigor and professionalism by fortifying the case report literature.

Unfortunately, the status quo points in the opposite direction. Most acupuncturists write only one or two case reports during their years of clinical training for the purpose of demonstrating their competence to graduate. The education of new acupuncturists in modern times involves listening to lectures and reading about theory, observing experienced practitioners, and participating in clinical internships. Practicing acupuncturists continue to gain knowledge through attending the occasional weekend lecture or training session given by experts in the field who discuss their own interpretations of theory and demonstrate special techniques. In this setting, cases are discussed, but oftentimes informally, without specifying important details, such as treatment duration or frequency, needle technique and retention time, and offering only the vaguest of descriptions of outcomes. The advent of social media and online forums have unfortunately further deteriorated the level of discourse, where many acupuncturists briefly recommend point protocols to each other without any diagnostic considerations whatsoever. The vast majority of practitioners forge ahead, doing their best to provide patient care but without pausing to record their experiences in written form to share their growing expertise.

Reversing these trends is an arduous task, and no one book can hope to stem the tide alone. As a necessary first step, however, this book methodically presents the process of conceiving and writing a detailed case report. Case reports can be used not only to support clinical trial research, but more importantly to describe in detail the thought processes behind how theory is applied to create effective treatments for individual patients. When experienced practitioners share their clinical pearls, all patients have the chance to benefit.

Both clinical and mechanistic biomedical research have no doubt brought us a long way. Without the tireless work of researchers during the last 50 years, it is entirely possible that acupuncture might have faded out completely due to implausibility of its effectiveness in the context of the current scientific world view. But fortunately, research has proven the efficacy of acupuncture in many areas, and the insertion of an acupuncture needle is seen as having validated physiological effects. At this point, the field is at a crossroads. With significant validation already in the literature, research is shifting in new directions that take traditional theory into account. Mechanistic research may enable acupuncturists to refine treatment choices and incorporate new technologies into the traditional paradigm. Clinical researchers are creating new types of studies where treatment is individualized to match the specific patient, rather than reducing acupuncture to a collection of protocols. Research can be directed in a way that respects the theoretical background of acupuncture; indeed, it is interesting to observe that acupuncture protocols of clinical trials are created by consulting with senior practitioners of traditional acupuncture.

However, if research is directed in a protocolized manner that simply tests the efficacy of acupuncture, practitioners of the future will carry it out as a procedure rather than as a paradigm. Traditional ways of training will fall out of favor, and traditional practitioners will be marginalized. If acupuncture continues to be practiced with little training or regard for traditional considerations, eventually there will be no one left who has a deep understanding of acupuncture to be able to design new protocols for future studies. Acupuncture will become an empty shell, and the well will run dry.

For it to survive, the field of acupuncture must balance the *yang* of heady research with the *yin* of actual clinical practice. The case report record can not only help us to better teach and learn acupuncture, but also can safeguard the transmission of the traditional perspective beyond the next generation. This genre of medical writing empowers experienced practitioners to communicate ideas on how to apply both ancient theory while incorporating new technologies and scientific knowledge. If the field aggregates case report literature, it may find common threads to support the advance and development of acupuncture in practice — innovating while preserving important traditional elements. While western science continues to investigate acupuncture's efficacy and mechanisms, those with a traditional theoretical outlook must not neglect case reports as an important branch of research if they wish to preserve the integrity and strengths of traditional acupuncture.

I invite you into the world of writing acupuncture case reports. We have much work to do.

Part I

Before delving into the nuts and bolts of how to write a case report, it is important to consider the status of and the reasons for writing case reports as they relate to medical literature. The first half of this book discusses the background of acupuncture case reports — their history, their role in science, and their influence on clinical practice. Chapter 1 describes the history of case reports, identifying differences between eastern and western medicine in their reasons for writing about cases, and clarifying the role of case reports within the context of evidence-based medicine. Chapter 2 then pivots to examine acupuncture as a traditional form of healing distinct from the materialist and reductionist perspective of biomedicine. Many styles of acupuncture exist, and the chapter investigates how case reports can help us to understand how to apply specific strategies in the treatment of individual patients. Chapter 3 discusses how the development of acupuncture therapy was not based on the western scientific method but rather on experience and careful observation of cases as a way to generate traditional theories. It notes the inherent difficulties in testing acupuncture's efficacy with the scientific method and describes how leaning too heavily on the biomechanisms of acupuncture rather than on patient care may be eroding its potential benefits. Chapter 4 recasts the practice of science as a human activity relying on skills that include making observations, inferences, and predictions. These skills are not only used in biomedical practice but in traditional acupuncture as well, and are essential for both pursuing knowledge — both through the scientific method (examining phenomena as they pertain to large populations) and through writing case reports (examining phenomena as they pertain to one individual). Finally, Chapter 5 discusses a practical conception of integrative medicine that brings together biomedicine and acupuncture. Rather than modeling integrative practice based on logistical, financial, and political concerns, using case reports to identify and consider the strengths and weaknesses of each type of medicine can help us to understand how to combine them in a way that benefits patient care.

CHAPTER 1

The History of Case Reports

A CASE REPORT is a clear, concise account of a single patient. For thousands of years, the interactions between physician and patient have been recorded in many forms, from brief sentences inscribed on ancient papyrus to formal peer-reviewed articles published in modern medical journals. In both acupuncture and biomedicine, case report records were kept for historical, educational, and research purposes, and they continue to be a valuable contribution to the medical literature. In the modern era of scientific medicine, physicians often think of the purpose of case reports as generating new ideas to stimulate research through clinical trials, yet many other valid reasons exist for writing case reports. For example, some case reports are written to educate the reader about an unusual pathology or to describe a new technique, whereas others are written to demonstrate good clinical reasoning or to point out pitfalls to avoid in practice.[1]

A common misconception about case reports is that they are anecdotal in nature. The term 'anecdote' has the connotation of being unreliable, often used in a disparaging way to suggest that case reports are not scientific.[2] In truth, as an account of an actual patient, the modern case report is a piece of scientific evidence, which includes detailed observations and specific data about the initial condition of the patient, the treatment, and the results as well as arguments based on the main points of the case. The fact that only one patient is described, without control over variables or patient blinding, may lead to criticism of it as a valid piece of evidence. However, sometimes a case report is the best evidence available as a basis

13

for a medical decision — especially given that many patients seeking alternative care have complex histories where the conclusions from randomized trials may not be applicable.[3] Case reporting can be thought of as a broad method of producing valid knowledge that uses a type of reasoning that differs from the inductive approach of the scientific method.[4] The advance of science is often viewed as being the result of experiments, but science also advances through other styles of reasoning, including deducing new ideas from past knowledge, using mental models, classifying, and making testable predictions. A physician performs all of these activities when treating an individual patient, and the case report format allows for clear communication on how these skills contribute to the practice of medicine.

With regard to terminology, most medical journals use the term 'case report' when considering a written account of a single case. Informally, the term 'case study' is often interchanged for case report, but technically case studies — as an examination of a particular individual instance within a greater context — can span a variety of fields that extend beyond the practice of medicine. Case studies in the legal field provide precedents to help future judges make decisions. Case studies in the economic field may help businesses find solutions to navigate complex and difficult situations. In the teaching profession, an instructor might introduce a hypothetical 'case study' analysis as an educational tool in order to test student knowledge and problem-solving skills. Because the idea of the case study can be applied to other fields, and in some instances can describe pedagogical exercises rather than actual patient interactions, the term 'case report' is a more precise term for a paper describing a single patient. At the same time, the purposes of a case report may have the potential to achieve goals similar to case studies by providing information that may assist practitioners in decision-making, helping them to find solutions to navigating difficult and complex situations, and demonstrating specific ways to benefit patients.

Case Reports in Western Medicine

In western medicine, the case report as a genre of medical writing has evolved over almost four millennia. The earliest medical case records, preserved on Egyptian papyrus, date back to about 1600 BCE. Classical Greek physicians who were contemporaries of Hippocrates (c. 400 BCE) provided objective and detailed descriptions of the findings and courses of various illnesses. In 200 CE, the Greek physician Galen famously described many of his cases. Towards the end of the first millennium, Islamic physicians Rhazes and Avicenna wrote detailed descriptions of individual patients and their illnesses, in one case differentiating small-pox from measles and in another relating hand numbness after a fall to an injury to specific vertebrae of the spine. All these writers recorded detailed observations and explained their inferences in a way that readers were able to learn from their expertise.[5]

In the second half of the 16th century, a new form of medical writing emerged in Europe. Collections of case histories called *Observationes* became more popular over the course of the 17th and 18th centuries.[2] Originally, local physicians wrote these case histories for self-promotion, describing their success in practice. The tone was conversational and subjective, and physicians wrote in dramatic ways to make their work seem more impressive and appealing.[5] By the 19th century, the emphasis on success in these collections shifted to a new focus: on the knowledge of disease.[6] By describing detailed observations of individual cases, physicians endeavored to advance knowledge, reporting objective clinical findings in technical terms. In the latter half of the 19th century, the structure of scientific papers also began to shift. Louis Pasteur and Robert Koch, in their papers describing early experiments on the germ theory of disease, wrote their observations in a format which organized the facts of their discovery in a number of sections for clarity. These sections included the Introduction, Methods, Results, and Discussion (today commonly referred to as IMRaD).[7] For case reports, this format is modified slightly to substitute the Methods section with a Case Presentation section. As science developed and discoveries were made, journals and scientific organizations began insisting on this general format for all papers, including case reports.

Starting in the early 20th century, case report writing recorded new developments in scientific knowledge. New diseases first described in case reports included Cushing's syndrome, erythroblastosis fetalis, Ebola virus infection, and toxic shock syndrome. Adverse effects of new medications were first publicized in case reports, including thalidomide-related birth defects and thrombosis due to oral contraceptives. Case reports also described new interventions for difficult diseases, including the first surgical ligation of patent ductus arteriosus, the first use of lithium to treat mania and the first heart transplant.[3] In 1985, the *Journal of the American Medical Association* reprinted what it deemed to be the most significant 51 papers affecting the practice of medicine during the previous 150 years; five of these were case reports.[8]

In the 1980s, evidence-based medicine ranked the variety of sources of scientific information into a hierarchy of evidence, with randomized controlled clinical trials placed as the gold standard at the top of the pyramid with case reports relegated to the bottom. As case reports were deemed lacking in significance, funds and efforts were prioritized elsewhere, and case report writing languished. Comparatively, from 1946 to 1976 the percentage of case reports in general medical journals was as high as 38%, whereas by the 1990s the percentage of case report articles had been reduced to 17% — and down to the low single digits in publications such as the *American Journal of Psychiatry* and *Archives of General Psychiatry*.[2] One possible reason for this change was the development of diagnostic technology and tools such as rating scales to enhance measurement[9] at the expense of observation. Despite psychiatry being a field where patient presentation tends to be unique from individual to individual, the editor of the *British Journal in Psychiatry* in 2003 stated "I hastened the demise of the case report to exclude what I see as psychiatric trivia."[10] A journal's impact factor is influenced by the number of citations it sees, and because randomized control trials, reviews, and meta-analyses are cited more often than case reports, publishing case reports may reduce the journal factor and therefore the prestige of a journal. Some journals went so far as to refuse to accept case reports altogether,[11] even though some notable case reports, such as the first report of an individual with AIDS, have been highly cited.[12]

In the early 2000s, criticism of evidence-based medicine began to surface. One reason for this was that a statistically analyzed clinical trial might prove efficacy for an intervention tested on homogeneous populations, but the intervention might be ineffective for individuals with a more complex clinical picture.[13] With this in mind, prestigious journals such as *The Lancet* and the *BMJ* began publishing case reports again, leading to an increase in case report writing. The advent of electronic publication has also added to the increase in published case reports, with currently over 30 online journals publishing case reports in various specialties. As of 2017, 1.77 million case reports were indexed in PubMed, the most prominent database for scientific articles.[3]

Case Reports in Chinese Medicine

Case writing in Chinese medicine has a similarly long history. In the Shang dynasty (1600–1046 BCE), brief records of royal illnesses, including the patient (usually the king), practitioner, date, disease, and sometimes the treatment were inscribed into oracle bones (either ox shoulder blades or tortoise shells used as divining instruments by treating them with fire).[14] By 175 BCE case histories were being recorded in the official government records. Sima Qian, the Grand Historian of the emperor, recorded 25 case histories by the royal physician Chunyu Yi, including details regarding symptoms, facial appearance, pulse readings, and evaluation of channels. The doctor's first task was to predict if the patient's illness was fatal. If there was a chance of survival, therapy was briefly described.[15]

In the first millennium, case histories were told in the context of biographies and literary works. In the 3rd century CE, numerous cases of the legendary surgeon and physician Hua Tuo were recorded in histories such as the *Sanguo Zhi* (Chronicle of the Three Kingdoms). The level of detail in these accounts is sparse and diagnostic reasoning is rarely mentioned, as cases were presented more for their literary effect than for medical purposes. For example, one case of an abdominal operation involved a patient who was told that surgery could prolong his life, but not by more than ten years. The record reports that he successfully underwent the surgery but died exactly ten years later.[14] By the 7th century CE, the focus of written cases started to move away from brief stories of famous physicians to more detailed accounts that explained the nature of illness, illustrated effective treatment, or supported a theoretical position. For example, in the *Beiji Qianjin Yaofang* (Emergency Prescriptions Worth a Thousand Gold Pieces), Sun Simiao discussed the progression of *xiao ke* (wasting-thirsting disorder, linked with the modern condition of diabetes) through describing the case concerning the illness and death of an official in 636 CE.[14] In Chinese medicine, cases were written describing acupuncture, Chinese herbal medicine, and other related therapies based on traditional theory.

Another source of case-based information that has survived for nearly a millennium is consultation records made by physicians. At first, these consultation records were kept private, although it was common practice in late Imperial China for a physician to provide for the patient a written summary of his or her diagnosis and analysis along with the recommended herbal prescription. Because knowledge in the field was secret and passed down only from generation to generation, educated people found it difficult to break into practice if their family was not a part of an established lineage. In the 11th century, Yan Jizhong, a minor official interested in the methods of a local pediatrician, Qian Yi (1032–1113), went so far as to gather case records from Qian Yi's patients and associates and with some editing and commentary published a volume of 23 cases. Conventions of secrecy helped protect the livelihoods of most physicians at the time, and for this reason, this type of case collection was interesting reading for both novices and experts in Chinese medicine. Qian vacillated about whether to transmit his knowledge to the general public, but in the end publishing these cases validated his authority and established his reputation as a founder of Chinese pediatrics.[15] In this way, the case record began to evolve from private to public record.

Collections were written in a variety of formats, from small clusters of case narratives reported about a physician by outsiders to case histories as examples in the context of medical treatises to the printing of collections of cases, each written and selected by a single physician and describing his or her work.[15] In the 12th and 13th centuries, while the collection and publication of case compilations flourished, the practice of Chinese medicine continued to evolve. While contending lineages of learning all had their roots in the classical literature, they favored the importance of certain therapeutic principles over others. For example, Zhang Congzheng maintained that the main causes of disease were Heat, Cold, Dampness, Dryness, Summerheat, and Fire, preferring to cure disease through sweating and purging. Li Gao maintained that inappropriate lifestyle and diet would cause a decline in vitality, which could be remedied by strengthening the Spleen and Stomach. To illustrate their viewpoints in their respective works, case histories were interspersed among the theoretical material, thereby lending credibility to their respective approaches. Sun Yikui (1522–1619), a scholar who had learned secret prescriptions from an unknown Daoist, wrote a cumulative autobiographical work of Chinese medicine case histories towards the end of his life that could be compared to a *wenji* (a compilation of literary works attributed to an individual scholar).[15]

In the early 16th century, the first guidelines were published for recording cases in Chinese medicine. Han Mao from Sichuan, in his 1522 text *Han Shi Yi Tong* (Mr. Han's Generalities on Medicine) discussed recording the patient's name, place, and date, together with the patient's presentation, which was based on looking, listening, smelling, palpating, and questioning.[i] The presentation would be followed by discussion on illness origin, and then treatment and eventual outcomes. Han Mao named this more comprehensive format '*yi'an*', where *yi* means medical and *an* roughly means 'case statement' (the same terminology as a legal case statement). The format suggested by Han Mao changed case reporting from free form notations to a series of data fields requiring a particular pattern of observation and recording. Similarly to legal cases, a collection of medical cases could be used to compare similar presentations, tease

---

i While today authors are prohibited from including any information that would allow for a reader to identify the patient (e.g., the patient's name), conventions at the time allowed for the inclusion of this type of information.

out key differences and their ramifications, and find patterns useful in resolving difficult problems.[15]

From 1500 to 1800, the writing of *yi'an* case collections increased in number and diversity of form. Some collections of cases focused on various themes, such as demonstrating the classification of diseases or taking a closer look at a specific condition. Another approach to writing case books involved illustrating a number of similar cases but then including one or two examples describing exceptions to the rule. Some practitioners even compiled a volume of their own cases to demonstrate their thinking patterns and perspectives. In 1531, a famous collection of 123 case histories was published in a collection called the *Shi Shan Yi'an* (Stone Mountain Medical Case Histories of Wang Ji). The cases were compiled by Wang Ji's disciple Chen Jue and collectively described his approach to medicine. In 1549, Jiang Guan published *Ming Yi Lei An* (Classified Case Statements of Eminent Physicians), an encyclopedic compilation of cases from the medical writings of Song, Yuan, and Ming physicians, organized into 195 disease classifications.[15] In 1572, Nie Shangheng, a physician in Fujian province, wrote a case book focused on smallpox, demonstrating how patients with the same disease might have different enough presentations to warrant individualized treatment. It provided a range of examples to demonstrate solutions for the variety of presentations a doctor could face in clinical practice.[15]

The case literature grew to become one of the main avenues for Chinese herbalists and acupuncturists to demonstrate clinical experience and medical knowledge. Some authors, wishing to promote opinions on how to analyze and approach cases, expressed their interpretations of theory in the context of individual cases.[15] In the late 18th century, the famous Shanghai physician Ding Ganren (1865–1926) published his own series of case records, changing the format to indicate that the records were to be used for didactic purposes. The patient's condition and pulse assessment were described, followed by the diagnosis and treatment details. After the outcomes were reported, a short rhyming statement was included to help with memorizing the lesson learned. In order to garner Chinese government support and preserve traditional practice,

He Lianchen (1861–1929) advocated for adopting an objective format of biomedical medical case histories.[15] Case reports gradually included more detail about treatment and outcomes, and discussion with analysis was added to demonstrate the thought processes behind diagnosis and treatment design. The contents of case reports developed as writers worked towards sharing a complete account, and by the late 20th century, the biomedical IMRaD approach became the standard for written case reporting in Chinese medicine journals as well.

The Purposes of Case Reports

The histories of case report writing in acupuncture and biomedicine have striking similarities. The precursors to modern case reports in both domains began over two millennia ago, and early cases were described with little detail or with missing components. As time went on, the specifics of cases were described more thoroughly, including the patient's presentation, diagnosis, treatment, and outcomes. Explanations discussing the significance of each case were added, and in recent years, a standard format for both acupuncture and biomedicine case reporting has evolved, ensuring that all relevant details are included.

Despite this, the purposes for writing case reports in biomedicine and acupuncture have diverged over time. Early on in both disciplines only cases involving the nobility were recorded, but over time other individuals became the subjects of cases, and the purpose of case reports expanded to writing for self-promotion. As time passed, case studies in both domains began to explore the nature of health and to record discoveries made through observation. But in the West, as the scientific method caught on as a major driver of medical knowledge, case reports came to be written as a pathway to generate hypotheses for larger trials. Influenced by the journal impact factor, publication of case reports in modern biomedical journals favored case reports detecting novelties, including unusual clinical findings and adverse events[16] as these are more likely to generate citations.

In contrast, acupuncture case reports were written for educational and practical purposes: to elucidate difficult theory, describe the specific skills and observations associated with diagnosis, and provide a rationale for treatment. Because traditional theory can neither be refuted nor supported by the scientific method, early cases did not seek to form hypotheses for scientific testing but were used more often for instructional purposes involving the transmission of experience. After acupuncture was brought to the West and viewed under the lens of biomedical mechanism and clinical trial, case reports in biomedical journals gravitated towards the now common conclusion of hypothesizing that 'acupuncture may be effective for condition x; more research is suggested.' Yet focusing case reports on making research hypotheses overlooks the important role of case reports in acupuncture education and practice. The types of traditional knowledge transmitted through writing and reading case reports in acupuncture were neglected, as the importance of unbiased scientific validation rose due to the advent of evidence-based medicine.

Evidence-Based Medicine and Case Reports

In the late 20th century, evidence-based medicine became the predominant model of medical care. Gordon Guyatt and colleagues introduced the term in 1992, defining evidence-based medicine as requiring "the integration of the best research evidence with our clinical expertise and our patient's unique values and circumstances."[17] The biomedical community quickly gravitated to focusing on the first half of the definition, with a hierarchical system of classifying evidence becoming the cornerstone of evidence-based medicine.[18] In this approach, evidence is "any data or information obtained through experience, observational research, or experimental work. This data or information must be relevant either to the understanding of the problem or to the clinical decisions, (diagnostic, therapeutic, or care oriented) made about the case."[18] "Best research evidence" was determined by basic medical sciences, from the accuracy of diagnostic tests and prognostic markers to the efficacy and safety testing of specific therapies. At the top-most level of the evidence hierarchy is the double-blind randomized control trial (RCT), viewed as the gold standard of evidence efficacy. RCTs test the success of a standardized protocol across a large group of subjects to ascertain how the group responds on average.

At the opposite end of the evidence hierarchy are case reports. Although they certainly qualify as a type of evidence, especially useful for conditions which have not been thoroughly researched, case reports have traditionally been relegated to the bottom of the hierarchy due to their potential for bias, unrepeatability, and their lack of statistical significance.[19] Case reports have often been dismissed as 'prescientific' and incapable of yielding valid conclusions about cause and effect relationships because of the presence of uncontrolled variables that might affect outcomes.[20] Conclusions from a single case report do not reveal the accuracy of diagnostic tests and prognostic markers nor can they serve as the basis for any generalized conclusions about clinical efficacy. In biomedicine, case reports have historically been written mainly for hypothesis building or for elucidating unusual aspects of patient care.

Yet the evidence hierarchy which evidence-based medicine relies on can be misleading in at least two major ways. First, clinical trials are designed to examine a narrow representative slice of the patient population in order to maximize the likelihood of detecting positive findings. However, the specificity of the group selected means that the evidence may not be applicable to many patients who might not fit within the criteria specified in the study.[21] For example, efficacy trials use inclusion and exclusion criteria to create a homogeneous patient population to study.[ii] Because the patients in an efficacy study are often limited by age range and other specific diagnostic parameters, the outcomes and conclusions of efficacy studies may not accurately reflect the overall effectiveness of an intervention across a larger population. For example, if a study focuses on individuals between 30 and 50 years old, the evidence might not be ideal for making a decision on what to recommend for a 70-year-old patient. For the atypical patient with a complex history who might be excluded from the study, guidance is limited and conclusions from clinical trials may not be as straightforward to apply.

---

ii Effectiveness trials, testing a more flexible approach to treatment, tend to have more heterogeneous patient population, which may be somewhat more useful as evidence in practice.[22]

If the individual has symptoms reflecting a particular condition but has unusual lab tests, the patient might not respond in the same way that a typical patient with that same condition might. Creating a study that allows for statistically validated treatment results requires that numerous categories of patients be left out; if many are left out, even though the research methodology may be robust, the evidence may have limited usefulness.

Second, the act of standardizing the intervention to be tested in a clinical trial will further limit the usefulness of the evidence. In biomedicine, a clinical trial testing the efficacy of a medication may test a specific dosage of a single agent on trial subjects. If successful, physicians will apply this evidence in clinical practice, but may need to vary the dosage depending on individual factors including the patient's age, weight, and diagnostic tests. If the dosage is too low, it can be increased to match the patient's needs. In an acupuncture clinical trial, the wide variability in the number of possible points and point combinations must be winnowed down to a point protocol. If the study proves acupuncture efficacy, the protocol chosen may be beneficial for the statistical average, but for a good number of patients at either end of the bell curve, it may be misdirected and unhelpful. Even if efficacy is proven, an acupuncture clinical trial will offer no guidance for providers who need to modify acupuncture treatment to suit the patient. For example, if a specific protocol has been proven efficacious for urinary incontinence, but the individual patient in clinic doesn't improve, it may not be the case that the patient does not respond to acupuncture but rather that the protocol did not match the patient. One cannot simply raise or lower the dosage of acupuncture to fit the patient; adjusting the treatment would require varying the points or point combinations, which is not something that can be captured in a clinical trial. Declaring that acupuncture is not effective for a patient simply because a trial protocol did not work would be akin to declaring that a patient is unresponsive to all pain medications after only trying ibuprofen. Other options for adjusting the treatment to suit the patient exist, yet the context of the clinical trial prevents flexible application. Adopting a blanket treatment protocol wholly without consideration for the patient does not conform to traditional principles, nor does it seem to satisfy the patient-centered mandate of evidence-based medicine. Adjusting the treatment points requires experience; evidence from case reports in this way may be instrumental in informing patient-centered care.

At the same time it is important to note that the rigor of case reports as evidence can be evaluated through several parameters: internal validity, construct validity, and external validity.[23] These parameters have been used in building theories in patient-centered disciplines such as psychology, where generalizations are difficult to make due to variations between individual cases. Internal validity describes the likelihood of a study to support a reliable cause and effect relationship. Internal validity tends to be low in case reporting, due to the many variables uncontrolled in the observation of a single case. However, using cross-case analysis to consider the data can help to increase the internal validity.[24] Construct validity — the degree to which an inference is considered logically sound — depends upon the ability of a test, or a group of measurements, to support an abstract construct. Using multiple objective sources of data from a variety of testing methods may more strongly validate the construct.[25] External validity, or generalizability, is the likelihood that the theories generated may apply to situations outside the original case.[23] While individual cases are not sufficient for making statistical generalizations about a population, aggregated case reports examined collectively may be a starting point for generalizing theory out of empirical observations. It has been suggested that between four to ten case reports may be a solid basis for analytical generalization.[24] Case reports individually may not have power, but rigorous analysis of aggregated case reports may support the theories underlying the paradigm of traditional acupuncture, in contrast to clinical trial research which tends to support a standardized version of acupuncture that does not take the patient's individual circumstances into account.

In sum, the clinical trial does not merit the pedestal that it has enjoyed for decades when considering the value of complementary therapies that are characterized by individualized treatment, while case reports have several properties that recommend their use. There is a difference in the type of knowledge that can be learned from the study of large population epidemiology and from the analysis of an individual case.[26,27] The strongest piece of evidence within the hierarchy of the evidence pyramid — which might support the use of a particular intervention over large populations — may not be the best piece of evidence for an individual patient when case reports are examined.

Evidence-Based Medicine and Acupuncture

The single-minded focus on RCTs as the 'best research evidence' therefore does not reflect the importance of evidence to be considered in light of clinical practice, which involves diagnosis and selecting treatment for an individual with specific characteristics. Unsurprisingly, testing acupuncture successfully with the gold standard of an RCT has proven quite difficult. From a research design standpoint, a variety of obstacles exist, including establishing a valid inert treatment to test for placebo as well as satisfying the debate between formulaic treatment protocols and individualized acupuncture reflecting real-world practice.[28] But that does not mean that acupuncture does not have valid evidence even by the standards of evidence-based medicine. Its approach seeks to synthesize the best evidence available; if convincing evidence is not available, or does not apply to the patient at hand, a therapy supported by a lower level of evidence can be used (evidence-based medicine does not even reject outright the use of unproven therapies). A clinician can rule out high-level evidence on the basis that it is not relevant to a patient.[29] Even if there is a standardized intervention that works well for 90% of patients, that means 10% of patients may not respond, requiring other options for care. Even if the treatment may be efficacious for a large homogeneous study population, there is no guarantee that it will also be effective for a specific patient.

Despite its place at the high point of the evidence pyramid, using clinical trial evidence is therefore not consistently useful in acupuncture. Missing from the single-minded focus on evidence from RCTs is the experience required by practitioners in deciding what evidence is appropriate and relevant. Even in scientific biomedicine it is incorrect to simply assume that positive outcomes will follow when the treatment is blindly applied if efficacy has been proven. Physicians rely on experience to correctly apply the knowledge, and this is best learned from cases. In biomedicine, knowledge of cases can enable physicians to draw on knowledge gained from larger research studies, informing correct application and effective patient care.[30] In traditional acupuncture practice, case reports serve a similar purpose; however, the knowledge base for making clinical decisions is not derived from the body of clinical trial research but rather from the body of traditional theory. Acupuncture case reports bridge the gap between theory and practice, taking knowledge gained from the textbook and explaining correct application towards effective patient care.

In fact, the definition of evidence-based medicine—"the integration of the best research evidence *with our clinical expertise and our patient's unique values and circumstances*"[17]—requires evidence to be combined with experience (a point the original evidence-based medicine group took pains to point out).[30] While efficacy trials might constitute the most valid evidence, their conclusions do not directly address the evidence-based medicine mandate to include the elements of practitioner expertise or patient individualization to the treatment. Case reports provide a different type of information and use a different thought process from clinical trials and are valuable in making the translation from theory and research into practice. Using judgment is essential to evidence-based medicine, and acupuncture case reports, not clinical trials, may offer details on how this expertise might be applied successfully.[26]

"Clinical expertise" comes from clinical skills and past experience. For each individual provider, clinical expertise is learned from numerous sources. For many centuries, the traditional manner of passing acupuncture skills and information from master to student was through the apprenticeship system, which included years of observing cases in clinic. In the current acupuncture educational system, a minimum number of internship hours is established by each state as required for licensure; in many entry-level and advanced programs, observation hours are also required where students learn from experienced practitioners. After starting practice, an acupuncturist's clinical skills and expertise continues to be built from experience by learning in retrospect from both the successful and unsuccessful clinical choices made in caring for one's past patients. Learning begins with observing cases treated by experienced mentors and continues through the analysis of one's cases and experiences. What is essential to recognize is that expertise can also be acquired by reading from the case documents of past and present masters who applied traditional theory successfully to their patients. In fact, whether the reader is a novice or an expert, reading case reports written by others can add considerably to a practitioner's knowledge and understanding. Furthermore, the process of writing case reports helps practitioners to more closely examine their clinical choices, refining their skills while sharing experience and wisdom with other practitioners.

The definition of evidence-based medicine also includes a focus on and concern for patient values and circumstances. Respecting patient values might involve offering treatment modalities such as acupuncture that are less invasive and more holistic. Some patients may choose to seek out acupuncture as a way to recover function and quality of life instead of resorting to possibly risky surgical procedures. Others suffering from side effects of medications may choose to explore other ways to manage their health conditions. Still others may resonate with the holistic nature of acupuncture, seeking treatment that integrates the mental and emotional aspects of health together with the physical. Taking into account patient circumstances involves making any health care decisions based on the patient's individual characteristics and situation, with considerations including (but not limited to) age, gender, vitality, severity of disease, and comorbidities. For conditions where the clinical trial research does not support variations in therapy according to a patient's individual situation, practitioners must rely upon their own clinical training or experience. But they can also rely on the wisdom and experience of other practitioners who employ traditional theory. Consulting the case report records may reveal instances where therapy has been successfully applied to an individual with similar characteristics. Reading case reports may also give readers ideas on how to tailor treatments for patients with less typical presentations.

Good medicine should be patient-centered by taking into account the individuality of the patient. Case reports by nature examine the individual patient, and by their nature may be considered more useful than clinical trials for fulfilling the patient-centered requirement for evidence-based medicine. The three parts of this approach to medicine are: 1) gaining knowledge based on mechanism and clinical trial (evidence), 2) applying the knowledge appropriately (experience), and 3) making sure the application is patient-centered. By taking into account the importance of clinical expertise along with patient values and circumstances, a full appreciation of evidence-based medicine softens the hard scientific perspective for which modern medicine has been often criticized for — specifically its narrow conception of what evidence has value and therefore its possibly limited usefulness for individual patients.[13] Another way to express the goals of evidence-based medicine is that it seeks to integrate evidence with experience in a way that honors the individual condition of the patient. In many respects it is unfortunate that the moniker "evidence-based medicine" was chosen because the name obscures the importance attached to the experience necessary to decide which information is most useful for an individual patient. One of its founders summed it up best when he said, "evidence alone does not make decisions."[30]

Conclusion

Good medicine is a compromise between the valuable evidence drawn from top-quality research and the clinical skills and sound clinical judgment learned from experience.[23] In traditional acupuncture, because the clinical trial evidence suffers from design problems and inconclusive results, other types of evidence become more important in informing decisions regarding patient care. While case reports may not enable clinicians to make broad generalizations about groups, they are more suitable for guiding practitioners through the choices to be made with individual patients. One of the key reasons then for writing case reports is to share clinical experience and explain one's thought processes in applying the theory to the problem at hand.

Reading and writing case reports enable practitioners to increase their clinical expertise, thus informing clinical decisions in the treatment of individual patients with unique concerns. The nature of the clinical trial does not account for the complexity of patient presentations in the clinic.[31] Because acupuncture is an individualized treatment and the theory involves many interweaving concepts, case reports are arguably more useful as evidence than randomized controlled trials in guiding acupuncture treatment in an individual case. While RCTs convince the public of the value of acupuncture and even lead to its addition to health plans and hospitals, clinical treatment decisions are made by individual practitioners using experience to decide which evidence to apply. The decision-making process is described best in the case report, and the choices made are validated by their effectiveness in individual cases. Because modern research studies that are well-accepted may not be exactly applicable or the best for individual patients, the field of acupuncture relies on experience, and case reports are the records for collective experience.

CHAPTER 2

The Practice of Traditional Acupuncture

THE ROOTS OF traditional acupuncture began over 2000 years ago, with the earliest written records of the practice of acupuncture found on silk scrolls excavated at *Mawangdui* in China.[1] These scrolls contain written descriptions of the locations and trajectories of a discrete number of acupuncture channels which form the basis for a communication and circulation network in the body. Early acupuncturists began the practice of stimulating points on those channels at the surface to affect the function of distant and deeper areas connected through the meridian network, with the goal of restoring and preserving health.

Over its lengthy history, the traditional practice of acupuncture has evolved considerably, spanning a wide variety of styles, techniques, and schools of thought.[2] Acupuncture is commonly regarded as referring to a procedure where needles are inserted into specific locations on the body, but some styles of acupuncture include the application of pressure or other instruments to these points without piercing the skin. Properly understood, acupuncture is a traditional paradigm that uses the stimulation of specific locations on the body for healing purposes.

Acupuncture Traditional Theory

Choosing effective acupuncture points for a patient requires a deep understanding of traditional theory, a collection of ideas that explain the nature of health and how acupuncture can be used to maintain it. The theoretical basis of acupuncture is one of the hallmarks of its practice, and knowledge of the principles described in the classical texts serves as the foundation of the paradigm. One of the earliest known classical texts describing acupuncture is the *Huang Di Nei Jing* (Yellow Emperor's Classic of Internal Medicine), believed to be completed around 220 BCE. It describes not only the names, locations, and uses for 135 acupuncture points as well as the trajectories of the 12 major channels but discusses in depth the theoretical basis for acupuncture.[3]

Over time, traditional theory evolved to explain the dynamics of disease and health and the observable effects of acupuncture. The most basic concept in traditional acupuncture that explains and predicts its specific effects is the concept of *qi*, which has roughly been defined as 'vital energy'. The dynamics and transformations of *qi* serve to model functions, qualities, and relationships which are not tangible but nonetheless observable. A second important concept to traditional acupuncture is the *yin/yang* dualism that describes a way of viewing contrasting yet complementary characteristics as essential to the balance of health and life. A third central idea to traditional acupuncture is that of the five phases, where the elements of metal, water, wood, fire, and earth in the macrocosm (the world/universe) reflect specific functions in the microcosm (the human body). A fourth important idea concerns the six levels, where processes within the body can be attributed to one of six types of movement, ranging from outwardly-expanding to inwardly-contracting. Choosing specific acupuncture points according to the current state of the patient can help to regulate and shift movements away from imbalance and disease. From one generation to the next, these ideas were developed and expanded upon; additional concepts including inter-organ and inter-meridian relationships were added to explain specific observations, creating a rich multi-layered understanding of health and illness.

The advent of modern scientific medicine has shifted the practice of acupuncture significantly. Rather than viewing acupuncture as a paradigm based on theoretical considerations, modern research tests the validity of medical acupuncture as a procedure. Mechanistic research seeks to understand the biological effect of needle insertion, and successful studies over the last fifty years have increased our biochemical explanations for the effects of acupuncture needling. At the same time, clinical research seeks to understand the efficacy of acupuncture for specific conditions, and increasing numbers of successful of clinical trials have led to increased recognition for acupuncture. Despite advances in both mechanism and clinical trial research, the ability to derive patient-centered treatment through these costly endeavors is no better off, as these continue to measure the effects and efficacy of acupuncture as a procedure without consideration for learning how to choose points that may be helpful for an individual presenting in clinic.

The traditional acupuncture paradigm bases the diagnosis on theory, which can be flexibly applied to create a treatment designed specifically for an individual patient's presentation. Defining acupuncture as a procedure rather than a paradigm strips away the underlying logic behind traditional diagnosis and treatment and does not allow for patient-centered treatment design. Furthermore, the modern medical version of acupuncture lacks a framework for predicting which points will be effective for treatment beyond repeating the protocols and points which have already been clinically tested. Tellingly, the protocols and points tested in scientific studies are generally chosen according to traditional knowledge; thus even medical acupuncture ultimately rests upon the traditional paradigm. As biomedical researchers attempt to more closely simulate actual acupuncture therapy, some leeway has been given in recently designed studies to add or substitute acupuncture points into trial protocols depending on the presentation of the individual trial patient. This is a fair compromise, more closely validating a more authentic practice of acupuncture. However, beyond potentially validating this type of flexible protocol to be efficacious, a clinical trial does not have the potential to advise point selection beyond what is recommended by traditional experts, and as such cannot improve acupuncture beyond its current state.

While modern sensibilities might discount traditional theory as being antiquated and unscientific, without this theory, medical acupuncturists would practice a limited number of protocols without the ability to adapt treatments to individual patients. One of the main strengths of a case report is to explain how to adapt treatments to individual patients. Without an underlying theoretical construct, medical acupuncture does not have a way to do so. Our current directions in research may continue to help validate acupuncture as a procedure, but at the cost of discarding the value of acupuncture as a system, limiting its practice severely and inhibiting its ability to be flexibly applied to many different situations. In order to bring acupuncture to greater numbers of people, research is necessary, but without understanding the roots of traditional acupuncture practice, much will be lost.

Acupuncture and Clinical Practice

The value, consistency, and applicability of traditional theory are tested through clinical practice. The practitioner's first task is to apply this theoretical knowledge to uncover the nature of the imbalance of the individual patient. Questions about the chief complaint are certainly asked, but other seemingly unrelated questions not common to biomedical thinking may also be important in making a traditional diagnosis, such as the specific location of a headache, sensations of cold and warmth, or the presence or absence of sweating. Palpation is an important aspect of diagnosis; this includes palpation of points and channels for irregularities including but not limited to local tension, knots and nodules, and lack of resilience. A number of pulse diagnosis systems exist that allow the practitioner to detect subtle quality gradations which can reveal the location of the imbalance. Visual observation of the patient may include aspects such as examination of tongue color and coating, posture and gait, venous congestion, and swelling and redness of injured areas. The practitioner's sense of smell and hearing are also used to detect abnormalities reflecting a diseased state. These diagnostic skills are used to a varying extent by each practitioner, depending on the nature of his or her training and the style of acupuncture practiced.

The goal of diagnosis is to discover the underlying imbalance giving rise to the pattern of signs and symptoms observable in the individual patient.

For example, an asthma patient who has difficulty inhaling, is pale, and tends to be sad emotionally, would have a different underlying imbalance and therefore be prescribed a different set of points from an asthma patient who has difficulty exhaling, a reddish face, and a thick tongue coating. A patient with temporal headache who has sensitivity at point LV 2 (*Xingjian*) and is easily angered would be treated with a different set of points from a patient with frontal headache concurrent with sinus congestion. Diagnosis in acupuncture is based not on biomedical tests, but on traditional ways of detecting an underlying imbalance to address.

Using traditional theory to understand the unique patient's condition is the key element to making a correct diagnosis. It is of course possible for the practitioner to make an incorrect diagnosis as a result of missing a piece of the diagnostic puzzle; it is also possible for the practitioner to base the diagnosis on a theoretical statement that may have less utility in the specific instance, or to misinterpret the theory due to translational confusion. For these reasons, it is essential to marry theory with clinical experience. Traditional theory by itself can be a useful perspective, but it lacks clinical precision. This is similar to the idea that in biomedicine evidence by itself cannot make decisions. While a variety of diagnostic methods may be learned in the classroom and from written textbooks, knowing what to pay attention to concerning an individual patient as well as how to interpret findings in relation to the appropriate theoretical statement is the domain of clinical education and case reports.

Once the traditional diagnosis has been made, the next step is to select and apply a corresponding treatment based on that diagnosis, using the traditional knowledge of the use of points to correct patterns of imbalance. A fundamental question to the study of acupuncture is how do we know which points and techniques to use for a given patient? In the 1980s and 1990s, the number of English language acupuncture texts were few. A number of these were manuals that listed conditions, followed by several points that could be effective in the treatment of the given condition.[4] Some called this approach to acupuncture the 'cookbook' approach. Giving a novice practitioner a list of points with which to produce a desired result is similar in analogy to giving a home cook a list of ingredients but without directions to produce a particular dish. It may be possible for a talented cook to approximate a reasonable rendition of a caramel apple tart given only the ingredients. But providing a novice simply a list of ingredients without adequate technical instruction on how to combine them would likely produce an unsatisfactory result. Similarly, with simply a list of points without rationale or training in needling technique, a novice acupuncturist might have a hard time achieving the desired beneficial effect. But a misunderstanding of how to make caramel by the novice cook only results in a ruined desert; the negative effects of misapplied theory and poor technique have far greater consequences for the overall health of a patient seeking treatment. Having only a list of points without the context of traditional theory leaves much to be desired.

Understanding why a point would be useful in a given situation is key to reaching an effective treatment. The use of the point depends on the clinical situation: imagine a text stating that a particular point is useful for treating constipation. While this point may be useful in some cases, it may not in others; its applicability for any patient would depend on the context and whether this point is appropriate for his or her traditional diagnostic pattern. In biomedicine, similar considerations apply. Acid reflux medication may be helpful for some patients with epigastric pain, but there may be many types of epigastric pain that are not due to acid reflux and therefore would not respond to acid reflux medication. For a patient with a frontal headache in the *yangming* region, GB 41 (*Zulinqi*) as a *shaoyang* point may not be very effective, even though headache is listed among its indications in many texts. Or if a patient has lower back pain and the available texts available suggest BL 40 (*Weizhong*), M-BW-26 (*Yaoyan*), SI 3 (*Houxi*), KD 3 (*Taixi*), Tung 22.05 *Lingku*, GV 4 (*Mingmen*), and the Nogier auricular lumbar point, how does a practitioner determine which point or point combination is the best for the individual patient at hand? Each practitioner's preference may be a reflection of his or her own teacher's experience and guidance, or the preference may depend on his or her own experience working with these points.

Acupuncture and Case Reports

There is an additional source of information practitioners can turn to for guidance. One way an acupuncturist can extend their own experience in order to make effective decisions is to read detailed case reports, which describe the reasons behind choices made in choosing points and in developing a treatment. Reading a full case report to understand the individual patient's situation would clarify whether the anticipated point usage would be appropriate and might be more likely to garner results.

As with the process of diagnosis, the selection of treatment points and combinations is also dependent on traditional theory, and in a similar fashion a practitioner may make errors in devising treatment. They may interpret theory incorrectly in light of the specific patient's condition, or they might base the treatment on a theoretical concept that may not be as useful in that particular instance. Even though traditional theory can provide a useful complex model of human health and healing, it is not sufficient to support practice without the guidance of clinical experience. And while direct observation of cases and discussion in clinical rounds led by experienced practitioners is essential to clinical training for students, the documentation involved in writing case reports can make these records more permanent and accessible not only to students but also to practicing acupuncturists looking to further hone their abilities. Writing a case report describing a successful treatment can help in translating traditional theory into actionable intelligence for future clinical settings.

The two-step process of using traditional theory to diagnose and then to develop treatment may in some cases be difficult to navigate. Leading a trainee through both correct diagnosis and effective treatment while demonstrating how theory informs the whole process is the essence of teaching acupuncture. Therefore, case reports can and should play a crucial role in guiding acupuncture education, and practitioners should make writing them a priority for furthering understanding of the field. Not only do they serve to help preserve traditional practice methods, but innovators in the field who focus on specific areas of diagnosis and

treatment can build on their experiences to move the field in new directions — even eventually evolving new styles of acupuncture.

Case reports can also be extremely helpful with regard to the many
different styles of acupuncture. A style of acupuncture is characterized
by a specific methodology of diagnosis and treatment. It can involve
the favoring of particular methods or theoretical concepts as a way to
diagnose an individual. A unique style can also specify technical aspects
about which points best address that diagnosis and how those points are
most effectively stimulated in treatment. One of the aspects that makes
acupuncture liberating to practice is the great variety of approaches;
however, at the same time, the number of possibilities available for treatment can be overwhelming. Case reports geared to a particular style can
offer clarity with regard to treatment options and outcomes.

For example, a common style of acupuncture is Traditional Chinese
Medicine (TCM) acupuncture, where practitioners assess the patient's
signs and symptoms by grouping them into patterns of disharmony. For
example, a patient with a chief complaint of insomnia would be diagnosed with one of at least eight likely patterns. With a symptom picture
that includes headache, dizziness, irritability, and a red tongue with a
dry yellow coat, he could be diagnosed with insomnia with a pattern of
Liver *qi* stagnation with Fire. A group of treatment points would then
be chosen to address this pattern, thus relieving the chief complaint of
difficulty sleeping, but also potentially having a positive holistic effect
relieving related symptoms of headache, dizziness, and irritability within
the context of the TCM pattern.

Japanese styles of acupuncture often use the five-phase *Nan Jing* (Classic of Difficult Issues) theory in choosing specific categories of points for
treatment. Pulse and abdominal palpation tend to be more frequently
used in diagnosis in Japanese styles, and their practitioners generally tend
towards a gentler needling technique. Practitioners in the Worsley Five-
Element Acupuncture style diagnose patients with a Causative Factor
by observing the patient's color (facial hue), odor, the sound of the voice,
and the way that emotions are expressed. Treatment points are chosen to
address that Causative Factor, sometimes according to the Chinese name

of the point. Case reports can clarify both diagnosis and treatment for readers and deepen their understanding of each style.

Orthopedic acupuncturists with a specific interest in treating musculoskeletal disease may diagnose by observing posture and gait, using orthopedic tests, and palpating the body. Treatment involves needling localized anatomically significant points including but not limited to trigger and motor points, combined with points used to move *qi* and Blood, unblock the sinew channels, and support any contributing pattern imbalances. Each of these examples of acupuncture styles tends toward particular modes of detecting the underlying cause of the patient's symptom or condition. Practitioners of this style also gravitate towards a subset of techniques and point choices — all of which can be better understood through reading case reports.

Case reports highlighting specific styles of acupuncture can also be useful beyond learning new diagnostic considerations or new rationales for point choices. A practitioner may become interested in seeking advanced training if the approach is intriguing, or she may refer a patient with similar presenting features to a practitioner who knows the style if special training is required for an advanced technique. Or, she may be able to learn finer aspects of correct diagnosis and of selecting effective treatment points. Case reports that involve mixing different styles may even discuss what each style contributes to the overall effect. And case reports which contrast the use of different styles at different points in a patient's treatment regimen may help the community to understand what kinds of patients respond best to specific styles.

Case Reports on Techniques

Beyond the variety of styles of diagnosis and treatment, the tools of the acupuncturist have also become more diverse. Case reports can help sort through the myriad of options available and their proper use and application. Even though the *Ling Shu* (Spiritual Pivot)[3] described nine types of ancient needles that covered a range of functions from 'fire' needling to surgery, in modern practice, only one of these needles (the filiform needle) is seen most commonly in practice. Modern manufacturing techniques have allowed for mass production of high-quality needles, each brand with its own tensile strength and feel. Gauges vary depending on the sensitivity of the point and the patient, and depths vary depending on the point and the style of acupuncture. Specialized manipulation techniques may also be used to achieve particular goals. For instance, small intradermal needles (0.3 mm to 6 mm in length) are meant to be embedded in the skin and retained for several hours to several days. A variety of needle types are used to prick the skin briefly to express a few drops of blood for certain types of conditions. A plum blossom needle, having a compound head with multiple points, may be tapped lightly on the skin to stimulate larger areas. Again, case reports can offer crucially important information to practitioners about which specific techniques enhance effectiveness best in what situations.

The practice of acupuncture continues to change and evolve. Even the translation of the Chinese word *zhenjiu* into the English word acupuncture has had ramifications; because the English word is literally derived from 'needle puncture,' one might assume that acupuncture requires a needle that punctures (the skin or the body). The Chinese word for acupuncture is actually a combination of two characters, with the first character representing the needle, and the second character meaning moxibustion, the application of heat to acupuncture points via the combustion of the prepared herb, Artemisia vulgaris. While burning moxa in conjunction with inserted acupuncture needles is common to a number of acupuncture styles, moxa stimulation is also frequently performed at acupuncture points without the presence of needles.

Beyond moxibustion, tools have been used for stimulation of points and meridians in traditional acupuncture practice. Needle stimulation without insertion is a technique common to acupuncturists in the Toyohari style, who use a teishin, a modified version of one of the nine ancient needles described in the *Nan Jing*, to carefully contact chosen acupuncture points without actually puncturing the skin.[5] A variety of handmade Japanese instruments are crafted out of silver, gold, and other metals to stimulate the points with gentle pressing, tapping, or stroking techniques without actually piercing the skin. A technique routinely taught in acupuncture courses, *guasha* is the use of a smooth instrument like a porcelain spoon or carved animal horn to scrape the skin over particular regions and pathways. Used to reduce surface stagnation at acupuncture points and regions, cupping involves the use of a glass or plastic cup with a vacuum created by a heat source or by a hand-held vacuum pump. Magnets may be taped onto the skin to activate acupuncture points, and mild electrical current may be applied with various frequencies, patterns, and strengths associated with specific physiological effects. Low energy lasers may be used to stimulate acupuncture points to good effect,[6] and even tuning forks with specific vibration frequencies may be used to stimulate acupuncture points.[7] Essential oils and oil blends may also be applied topically on points and regions,[8] and Chinese herbal extracts or other biologically active compounds may even be injected into the body at specific acupuncture points. Advances in research and our growing understanding of human physiology can feed further development in the evolution of acupuncture but understanding how to apply the range of ideas and techniques to individual patients will depend upon case reports.

Over the years, many currents of acupuncture have developed with a wide variety of diagnostic lenses and rationales for choosing treatment.[iii] New styles and methods have been able to flourish as individual practitioners innovate by reinterpreting and finding new applications for theories. Some make their mark by discovering new points, while still others find new ways to stimulate those points. All of these developments are based on clinical experience, testing new techniques, and finding out what works in the context of clinical cases. Yet these advances are not widely disseminated because while practitioners test and formulate new ideas for treatment, they tend not to record their discoveries in case writing. When new techniques gain consistent results in clinic, writing case reports on these new techniques that document the practitioner's experience is essential

---

iii For more information on the more commonly practiced styles of acupuncture, please see Appendix A.

to enable sharing these findings with the greater acupuncture community and allowing for further development of the field.

The diversity of currents offers acupuncturists multiple perspectives to consider and solve problems. At the same time, the diversity of practice styles makes it difficult for practitioners to know which diagnostic lens or which perspective on treatment may be the most useful for a given patient. As acupuncture practice methods continue to diverge, clinical practice can lose its focus, with acupuncturists applying multiple styles without really knowing what works best. Research efforts are also made much more complex by the existence of multiple styles, as experimental variables including needle stimulation, depth of insertion, and point selection become even more numerous and difficult to account for. In education, acupuncture colleges struggle to design their programs to teach at an appropriate depth of study while trying to include a broad exposure to the variety of approaches. While groundbreaking innovation is exciting, at some level the community of acupuncturists must find consensus on organizing the diverging currents so the field can have sufficient coherence to be accepted as a part of the modern healthcare landscape.

Case Reports on Choosing Effective Points

One final benefit found in documenting case reports is their use as evidence to start a discussion on how to frame acupuncture as a coherent modality rather than as a growing number of divergent and discrete currents of practice. Case reports can be helpful in a practical sense to recognize what each current contributes to the whole, while at the same time enabling us to understand the strengths of each approach. Case reports can assist in enabling us to better appreciate the common underlying theory of traditional acupuncture and use it to find points of connection among its divergent currents. Noting the convergent aspects among the various styles may help us to redefine acupuncture as a whole and its place in modern health care.

The body of traditional theory underlying diagnosis is common to all styles of acupuncture, though many styles tend to lean more heavily on a subset of the overall theoretical landscape when formulating a diagnosis. Treatments may have a fairly homogeneous needling technique within a given style, but needling techniques can vary significantly depending on the style. While the selection of points seems at first glance to be dictated by the originator of the style or the lineage holders, there are convergences in the way that acupuncture points are chosen for treatment across the various styles.

Local Points

One rationale for choosing a point in a given situation is because it is at or near the location of the patient's complaint. In TCM acupuncture, this local point approach is fairly common: LI 15 (*Jianyu*), TW 14 (*Jianliao*) and SI 11 (*Tianzhong*) are listed as good for shoulder pain, and they are unsurprisingly located on the shoulder. LU 7 (*Lieque*), located on the wrist, is useful for wrist pain. M-BW-24 *Yaoyan*, located on the lower back, can be helpful for treating pain in that region. But the use of local points goes beyond simply treating pain: BL 1 (*Jingming*), at the inner canthus, brightens the eye, and CV 17 (*Shanzhong*), at the center of the sternum, treats fullness of the chest.[9] Other styles use this same principle: Shudo Denmei uses GB 2 (*Tinghui*) in front of the ear for tinnitus[10] and Kiiko Matsumoto stimulates GV 20 (*Baihui*) on the crown with moxa for stagnation of Blood in the head.[11] One common principle among these and other approaches is that points may be chosen to affect the local region, either relieving symptoms by coursing *qi* and moving Blood in cases of excess, or through drawing *qi* and Blood to the region when levels are deficient.

The front *mu* (alarm) points and back *shu* (associated) points could be considered as local points for internal organ disorders. Front *mu* (alarm) points, a group of diagnostic points generally located on the anterior trunk, where tension or tenderness indicates an imbalance in the organ, tend to be on the surface of the body where the organ is located: the Bladder front *mu* point is located near the anatomical bladder, just above the pubic symphysis, and the Kidney front *mu* point is located near the anatomical kidney, just below the tip of the 12th rib.[9] Each of the front *mu* points also has indications where treatment of the point would benefit the related organ. The back *shu* points can also be considered as an extension of this idea, where dorsal points adjacent to the spine have been associated classically with the function of specific organs. Our modern understanding of the nervous system confirms the nerve roots from the spine enervate their respective organs.[12] *Ashi* points are a category characterized by sensitivity with pressure. Many of these points are local in nature. Motor points and trigger points as one type of local *ashi* point are found in individual muscles, and needling them can relieve local stagnation and ease pain. Some of these *ashi* points may be related to imbalances in related organs nearby.

Case reports can help us to refine our understanding of when to use an individual local point clinically. A case report may highlight a specific function and make the reader aware of situations where a specific local point may be useful. A case report may propose a potential mechanism for local point effects, depending on the detail of outcome measures and biomedical evaluation. A case report may even possibly recommend a particular style of stimulation for local points; for example, a patient who felt little relief after being treated with standard needling but experienced significant relief after electroacupuncture treatment on the same local points would be a good subject for a case report.

The use of local points is common to many styles of acupuncture. Case reports can be used to describe variations in locating these points as well as instructions on needle angle, depth, or technique that may be followed to achieve specific effects as required by individual patient presentations. They may also include guidance on which local points may be more useful for patients, according to traditional diagnosis and in light of individual characteristics. It is important to remember when reading a case report that the points chosen for the case report may or may not be appropriate for a different patient, so certain considerations need to be made when deciding whether or not this piece of evidence fits the patient, and how to adapt the usage or choice of the points if necessary.

The idea of using these local pressure points diagnostically reflects one of the hallmarks of traditional theory — to detect and treat the disease before it appears.[3] It is notable that some of the connections between these regions, although originally determined through intuition and observation, have been confirmed by mechanism. While the actions and indications of some acupuncture points may be explained or at least made more plausible to the western mind through a mechanistic rationale, therapeutic points were originally conceived and developed from observation of many patients. For example, patients with Lung symptoms tend to develop a specific sore and tender point at LU 1 (*Zhongfu*), inferior to the coracoid process on the chest just over the anatomical lung.

Channel Points

While local points have historically been stimulated or needled to benefit health, some traditions find that treating distant points may be preferable, either by having a stronger effect or being less risky to needling the local area itself. A second type of rationale for choosing a particular acupuncture point is due to the effect that it has on its channel or meridian. Traditionally, a channel is a pathway where the *qi* and Blood circulate. They can also be thought of as zones of influence, where stimulating points on a given channel affect the function of regions traversed by that channel. The twelve primary channels, each associated with an organ, include points on an extremity, along a limb, and by way of the abdomen connects with its associated organ.[9] In TCM style acupuncture, Lung channel points treat Lung pathology: to resolve Phlegm in the Lung, LU 5 (*Chize*) can be stimulated, while to clear Heat from the Lung, LU 10 (*Yuji*) can be needled. Because the Gall Bladder channel traverses the side of the head, one of the best TCM points for temporal headache is GB 41 (*Zulinqi*) on the same meridian at the foot. Points can be palpated for diagnostic purposes: in Kiiko Matsumoto style of acupuncture, the adrenal imbalance is diagnosed if pressure pain or tension is found at an alternate location of KD 16 (*Huangshu*). To relieve this tension, Kidney channel points are selected, commonly KD 6 (*Zhaohai*) and KD 27 (*Shufu*).[14]

The 12 primary channels have branches and associated regions called secondary vessels, including sinew channels, divergent channels, and *luo* channels. Points may be chosen on the primary channels to affect the function and flow of the secondary vessels. Eight extraordinary vessels also traverse the body and have more overarching effects on body pathology. A small group of distant points have effects on extraordinary vessels; for example, GB 41 (*Zulinqi*) may be chosen as a master point to regulate the *qi* in the Dai Mai, even though it is not technically on the Dai Mai itself; TW 5 (*Waiguan*) as its coupled point related by channel may also be needled to enhance the effect.[9]

Needling tender *ashi* points on an acupuncture channel can also sometimes have effects on distant regions within their zones of influence. Motor and trigger points are associated with pain patterns that spread beyond the borders of their local muscle to a related referral region.[13] For example, needling the *ashi* motor point of the flexor carpi ulnaris correctly on the forearm (close to SI 7 *Zhizheng*) causes the levator scapula insertion region near SI 13 (*Quyuan*) to release reflexively.[15] While some of the pain referral patterns are spread diffusely along the local area, it is worth noting that a number of the mapped referral patterns follow traditional acupuncture channels.

Distant channel points are often needled to resolve problems elsewhere on the same channel; however, channel points can also be useful in resolving problems on closely related channels. Each channel has multiple relationships that offer options for therapeutic solutions. For example, the point GB 34 (*Yanglingquan*) is known for its ability to spread the Liver *qi*. Although GB 34 is not on the Liver channel, this action is possible through the interior/exterior (*biao-li*) relationship between the Gall Bladder channel and the Liver channel.[9] A number of channel points have also been designated as crossing points; for example, needling SP 6 (*Sanyinjiao*), a crossing point for the Liver, Spleen, and Kidney channels, may have a therapeutic influence on all three.

A different type of paired channel relationship is the opposite clock correspondence, where for example the Bladder and Lung channels are related because the Bladder channel *qi* is most active between 3 and 5 pm and the Lung channel *qi* between 3 and 5 am. The polarity of their positions on the clock allows a practitioner to predictably locate a treatment point on the Lung channel that will effectively relieve a localized pain on the Bladder channel. Several other channel relationships exist, and while these relationships are the foundation for explaining point usage in the Master Tung style of acupuncture as well as styles based on *I Ching* theory such as the Balance Method, these theoretical relationships can explain the actions and indications of a good number of points whose uses were presumed to be merely empirical.[16]

Generations of acupuncturists have not only become skilled at finding points but also at making inferences about relationships between points. Often their discoveries relied on intuition and observations — for example, noticing that needling left side LV 4 (*Zhongfeng*) will normalize an abdominal reflex point at the left ST 27 (*Daju*),[14] or that pressing on a source point causes a change in the associated pulse position. By noticing in real-time with individual patients the dynamic nature of changes in the body, the ancients were carrying out their own version of experimentation as they tried to understand the nature of a phenomenon by performing the same safe procedure on many different patients, observing carefully and compiling one's own experience with that of others. Some early treatment choices may have been based on intuition, but over time, the accumulation of practical experience with the points led to the theoretical model of the channel system.

The connections between specific distal points and their zones of influence were originally discovered by experienced practitioners observing cases. Novices traditionally learn of these connections during an apprenticeship, and in modern times, this learning occurs through mentoring and educational programs. Sharing of these connections can be greatly enhanced by documenting them in writing. While a theoretical description may help, an actual demonstration of how this knowledge is used in the context of a case report has the potential to make the usage of these connections much easier to know when and how to apply. Specific areas where theory is debatable or unclear may be illuminated through case reports. For example, a case report may demonstrate the effectiveness of and reasoning behind a particular distal point strategy for an individual patient. Another case report might highlight how channel relationships are chosen to determine point selection in the Balance Method. For many graduates of a TCM acupuncture program, a theoretical explanation of the divergent channels is memorized but not commonly used in clinical practice; understanding in what situations and how divergent channel treatment is performed would benefit greatly by actual examples. Case reports demonstrating effective use of branching channels, *luo* channels, extraordinary trajectories, and other channel phenomena would also benefit the practice of traditional acupuncture.

Point Categories

Besides local and channel properties, acupuncture points are sometimes classified according to their inclusion in point categories. The points in these categories have common functional characteristics. For example, twelve points in a group called the *jing*-well points, located at the distal ends of each of twelve regular meridians, have common functions of clearing Heat and restoring consciousness. Any one of the *jing*-well points may be used because of these shared characteristics; for example, if a patient has a traditional diagnosis of Heart Fire and has lost consciousness, the *jing*-well point on the Heart channel HT 9 (*Shaoshang*) could be a useful point. Each channel has five transporting points (*jing*-well, *ying*-spring, *shu*-stream, *jing*-river, and *he*-sea), and each grouping has common actions. Each of the transporting points is also assigned one of the five phases (Metal, Water, Wood, Fire, and Earth categories), and a given point can be chosen for use in treatment due to its phase correspondence (e.g, LU 10 *Yuji* may be used because it is the fire point on the Lung channel). Numerous other categories for points exist, including those which have channel influences such as *luo* points, *yuan*-source points, and *xi*-cleft points, and those which have more global influences such as Celestial Windows (Windows of the Sky), Ghost points, *hui*-meeting points, Command points, Four Seas points, and lower *he*-sea points.[9]

The point categories are discussed in the classical literature in theory, but clinical experience with these points may help us to understand in which situations they might be more useful. For example, Sun Simiao's thirteen ghost points were listed in the *Qian Jin Yao Fang* (Thousand

Ducat Formulas) as being useful for mania disorder and epilepsy, but the specific usage of these points has been ambiguous. A case report which describes a specific presentation, explaining which points within this category were chosen and why, including needling style as well outcomes, would increase our understanding of this category and make it more useful as a result. Any category could be better understood through specific cases explaining their use in a case report, either by demonstrating the process of selecting a point because of its inclusion in a given category or discussing the specifics of how to treat a point in a specific category (e.g., when to consider bleeding a *jing*-well point).

Ashi points

Acupuncture treatment of tender *ashi* points has been a useful technique for thousands of years. Although the first record of using painful points as treatment points is found in chapter 13 in the *Ling Shu* (Spiritual Pivot), *ashi* points were actually first described in writing by Sun Simiao in the 7th-century work *Qian Jin Yao Fang* (Thousand Ducat Formulas).[1] Like all forms of acupuncture, simply treating the pain location without considering the underlying theoretical reasons for the pain is not *ashi* acupuncture. The cause of the pain is not necessarily at the location of the patient's chief complaint, and the *ashi* point related to the cause is often unknown to the patient until palpated. This idea is similar to other observations in traditional acupuncture theory where the cause of the problem is often not where the pain is manifested. *Ashi* points are arrived at in a number of different ways; some *ashi* points may be used diagnostically, where tenderness on palpation indicates a specific organ imbalance. Tender points can also sometimes be needled to rectify that organ imbalance. While some sources maintain that motor points and trigger points, as understood by modern muscle anatomy, are *ashi* points,[17] many traditional *ashi* points are diagnostic indications of specific imbalance elsewhere. For example, sensitivity at M-LE-23 *Dannangxue* on the lateral calf is indicative of Gall Bladder organ dysfunction.[9]

A case report can bring to light a new *ashi* point by suggesting when and how it might be used based on the details of the case. Alternatively, an *ashi* point already in common use may turn out to have unexpected results in a particular case. Noting the sensitivity of diagnostic *ashi* points may enable case report readers to detect symptoms before they occur. A case report may offer a helpful explanation of the location and needling technique for a specific motor point. Case reports are often notable for presenting new information, and as *ashi* points are often found outside the traditional channel system, these points are ripe for discussion.

Holographic Mapping

Beyond local points, channel points, point categories, and *ashi* points, a fifth type of point selection is based on holographic representations of the body. In traditional theory, the human is considered a representative microcosm of the earth. Channels are akin to rivers and the functions of the organs correspond to the natural elements of metal, water, wood, fire, and metal. As the microcosm (the human) can be understood through creating correspondences with a macrocosm (the earth), a smaller region of the body may provide a representation of the whole body. Several traditional diagnostic methods based on holographic representations have developed from collective observation and modeling; for example, radial pulse diagnosis at *cun, guan,* and *chi* positions correspond to the concept of the upper, middle, and lower *jiao* regions. Tongue diagnosis and facial diagnosis maps show correspondences between regions of the tongue and specific organs.[18]

Although the body was considered a microcosm with regards to diagnostic strategies in early Chinese medicine, it was not until the 20th century that the microcosm theory was applied as a treatment strategy. Numerous systems of holographic mapping have been observed and documented. Auricular acupuncture maps reveal hundreds of points placed on a representation of an inverted fetus on the external ear. Scalp acupuncture has several forms, the earliest of which located effective points based on their relationship to biomedical brain mapping. A system of abdominal acupuncture maps a diagram of a turtle superimposed onto the abdomen, where therapeutic points are chosen based on the location of a point on the turtle. The second metacarpal bone has also been used as a microsystem, with the region at LI 3 (*Sanjian*) roughly corresponding to the upper *jiao* (chest region), LI 4 (*Hegu*) to the middle *jiao* (abdominal region), and Tung 22.05 *Lingku* to the lower *jiao*. Up to 15 different zones within this small region of the body have been mapped out.[19]

The theory of 'as above, so below,'—reflected in the statement in *Ling Shu* (Spiritual Pivot) that "when the disease is above, select points from below, and when the disease is below, select points from above"[3]—inspired another way to consider mapping the body. The Balance Method, made popular by Richard Teh-Fu Tan, uses an anatomical strategy to find effective treatment points. Superimposing an image of the arm on the leg reveals structural similarities between the shoulder and hip (ball and socket joint), the elbow and knee (hinge joint), the wrist and ankle (gliding joint) as well as fingers and toes (phalangeal joints). Through recognizing the similarities between these structures, one can find effective treatment points; for example, points on the knee can be used to treat the elbow (and vice versa); likewise, points on the wrist can be used to relieve ankle pain. Choosing the most effective points with this strategy involves combining this mapping technique with channel theory: a pain at the knee on the Foot *jueyin* Liver channel, for example, might be treated successfully with a needle at a tender point near the elbow on the Hand *jueyin* Pericardium channel. Depending on the case, this process of mapping can be extended to finding effective treatment points by inverting the image, superimposing it on other body regions, and/or expanding or contracting the image.[16]

Internal symptoms can also be addressed through point selection based on this combination of holographic mapping and channel theory. Master Tung's point 88.01 *Tongguan* is located on the anterior thigh in the *yangming* region close to the Stomach channel. If one maps the torso onto the lower limb, the center of the anterior thigh corresponds to the chest region, and the relationship of clock opposites brings the Stomach channel (7 to 9 am) into juxtaposition with the Pericardium channel (7 to 9 pm), explaining how the point can be effective at treating heart palpitations.[20]

The number of possible holographic representations may seem overwhelming at first, leading practitioners to question which of the many possibilities will be most useful or effective. Case reports can help communicate to others what hologram(s) have worked well for a single patient; at the very least, one may be able to imitate the treatment for a patient with a similar presentation. Compiling case reports can help us to understand which systems may be more useful for which conditions. One challenge to using multiple holographic representations simultaneously is that understanding which system is responsible for the results is unclear. In this regard case reports which describe simple, elegant, and directed treatments may be particularly illuminating.

Empirical points

Finally, there is a category of points whose actions and indications seem to initially defy explanation. Empirical points are defined as points whose effectiveness has been proven by experience. While the actions and indications of many channel points are related to their associated organs, a few have actions and indications whose explanations are not as straightforward and may have been merely empirically noted in the past. However, through considering explanations of point indications that combine local, channel, and holographic images, it may now be possible to explain what was before merely an empirical observation. For example, LU 7 (*Lieque*) is the Command point of the Head and Neck, but since the Lung channel does not traverse the head or neck, one might assume that this point indication is empirical. However, if we consider holographic theory and superimpose the arm onto the torso, where the head represents the hand, the neck represents the wrist, and the elbow represents the navel, LU 7's location on the wrist corresponds to the neck. Then its relationship to the Lung channel via the Large Intestine channel through the interior/exterior (*biao-li*) relationship is explained, as is its connection to the Bladder channel through the opposite clock channel relationship.[21] Many points that were once considered empirical can now be explained through advanced theory.

Because many of the miscellaneous and extra points are not on channels, one might presume that they are likely to be purely empirical. However, as seen above in the case of LU 7 (*Lieque*) as a channel point, this is not always true. Other examples abound, such as M-UE-24 *Luozhen*. Used for a stiff and sore neck, the point itself is not located on a channel which traverses the neck; however, its location in a holographic image which places the neck at the level of the hand just distal to the metacarpal bases explains why Luo Zhen, SI 3 (*Houxi*), LI 3 (*Sanjian*), and TW 3 (*Zhongzhu*) all have neck indications. Or consider Tung point 77.22 *Cesanli*. It is not technically on a main channel, but its location halfway between the Foot *yangming* Stomach channel and the Foot *shaoyang* GB channel makes it a useful distal point for diseases affecting both *yangming* and *shaoyang* channel zones.[20] Presenting a case report on a new theoretical explanation as to why a particular point use or point combination is effective would be an appropriate way to publicize such findings.

One additional source of information which influences empirical uses of an acupuncture point is its traditional Chinese name. Before acupuncture came to the West, points were designated by name, rather than by their modern international code. These names, many of which were noted in ancient texts, generally consist of two (sometimes three) Chinese characters that reflect the function and/or other relevant qualities of the point. For example, PC 6 is the sixth point on the Pericardium meridian, but its Chinese name, *Neiguan*, translates as 'inner gate', reflecting one of its functions as a gate to access the inner landscape, treated to balance the heart and mind.[22] Kiiko Matsumoto has analyzed the written Chinese characters of a number of points to derive specific actions and indications, and in Five Element Acupuncture, point names translated into English carry particular significance in the context of their use to affect the mental/emotional sphere (i.e., 'spirit of the points').

Combinations of points

Whether the rationale for choosing treatment points relies on local, distant, categorical or holographic reasoning, most treatments consist of more than just one acupuncture point. Acupuncture points may be used together in combinations for a synergistic effect. In TCM acupuncture, several types of combinations are common, and a number of these were recorded in classical texts such as the *Zhen Jiu Da Cheng*

(Grand Compendium of Acupuncture and Moxibustion). One example involves sweating: for too much, reduce LI 4 (*Hegu*) and supplement KD 7 (*Fuliu*), while for too little, supplement LI 4 (*Hegu*) and reduce KD 7 (*Fuliu*). Other combinations are derived from theoretical statements, such as the combination of TW 5 (*Waiguan*) and GB 41 (*Zulinqi*) as master and couple points for the *Yang Wei Mai* extraordinary vessel. Points can also be combined in the 'chain and lock' combination, where several points at various intervals along the same meridian are needled in combination — for example LI 15 (*Jianyu*), LI 11 (*Quchi*), and LI 4 (*Hegu*) to treat cases of atrophy and hemiplegia. Another point pairing, which in theory reduces an excess from one point and shunts it to another point where there is a deficiency, is the use of the *luo*-source point pairs. For example, BL 58 (*Feiyang*), the *luo* point of the Bladder channel, used to reduce excess in the Bladder channel, can be balanced with needling KD 3 (*Taixi*) to supplement deficiency as the *yuan*-source point of the Kidney channel.[9]

Point combinations and prescriptions abound in non-TCM approaches as well. Specific combinations of four *shu* transport points on the distal limbs, selected according to the category and phase, are a hallmark of Sa Am acupuncture.[23] Master Tung acupuncturists will often combine two or three points in close proximity to intensify the effect or cover more of the target area in treatment.[21] The husband-wife imbalance in Five-Element Acupuncture calls for the combined use of four specific points based on five-phase relationships to resolve a treatment block.[23] In certain Balance Method strategies, combining four points in four regions — *yin* points on the upper left, *yang* points on the upper right, *yin* points on the lower right, and *yang* points on the lower left — creates a dynamic synergistic balance. These are but a few examples of how acupuncture points can be combined in treatment in a fixed or flexible manner. The overall arrangement of points in a prescription may also be shaped by the ultimate goal of the treatment.[24]

Understanding how to formulate an elegant and effective treatment is one of the challenges facing both new and experienced acupuncturists. Textbooks often will list a variety of point options for each pattern, but the process of selecting and combining points for an effective treatment is a skill that comes with observation and experience, and even the most competent practitioner can run up against unique cases. Reading case reports which explain successful treatment architecture that take into account the synergies between points may raise the level of sophistication and effectiveness of a new practitioner and offer insights for experienced acupuncturists.

Conclusion

Successful medicine depends on making good clinical decisions, and the driver behind decision-making should be what actually is successful in clinical practice. The best way to record and describe how clinical decisions are made is the case report. Clinical decisions are dependent on the use of traditional diagnostic skills, which in turn are reliant on traditional theory to help explain the individual patient's condition. Treatment point selection and choice of technique are also based on traditional theory, and the case report format lends itself well to conveying the thought processes behind the application of that theory to generate effective treatment.

Over the years, case reports were recorded and used as examples for instructing practitioners on the use of particular diagnostic skills, treatment techniques, or theoretical application. Reading published case reports can add to collective experience, allowing practitioners to expand their knowledge by understanding what has worked for others. Case reports may help us to understand the broad principles of point selection (local points, distal/channel points, point categories, holographic principles, empirical concerns, and point combinations). Within each grouping case reports can even further refine our usage depending on the situation, and discussion of case reports can shed light on the capabilities of acupuncture and when to use specific styles for diagnosis and treatment. The acupuncture case report literature is therefore essential as a guide for the acupuncture profession as it expands and diversifies its theoretical approaches and practical applications as well as preserves traditional ideas in light of its continuing evolution.

CHAPTER 3

The Science of Acupuncture

Our study of science starts in the early years, as children in grade school planting seeds in plastic cups and watching them grow. Middle school students pore over science textbooks replete with photos of organisms classified into groups, electron microscope images of living cells, and artist renditions of internal organ systems. High school students perform simple experiments and memorize thousands of facts about the physical and natural world. This education aims to provide us with the background knowledge necessary to be able to understand the nature of the latest discoveries. Mainstream media generates daily reports on new scientific findings in the natural world — about astronomers revealing new images of Pluto or entomologists identifying a new type of spider in Australia. Discoveries about our natural world abound, with researchers at the forefront of their respective scientific fields eager to share their latest findings with the world.

While science has allowed human civilization to reach new heights, the word 'science' has been overused to the point where its exact meaning is no longer clear. The parent of a young child might view science as an appreciation for nature, whereas for the student, science might instead represent a series of facts and equations that need to be memorized for the next test. For the average media consumer, science might be viewed as an understanding of mechanisms that explain how the natural world works, while for a hospital administrator might interpret it as the use of the most recent clinical trials to determine the best treatment for a condition. This wide variety of perspectives on the definition of science has value in generating a diversity of entry points, yet at the same time a lack of unanimity can lead to a lack of consistency in the criteria applied when approving specific medical interventions.

Science and Biomedicine

Despite what many people think, the use of science to determine what interventions are acceptable in medicine is a fairly recent phenomenon. It was not until 1910 that science became the driving force behind the practice of medicine. An American educator named Abraham Flexner, with the help of a circle of prominent physicians, published a report criticizing medical schools that included approaches to medicine regarded as unscientific. As an educator, Flexner traveled across the US and Canada, observing the training of physicians and evaluating medical schools.[1] A number of schools such as Johns Hopkins were evaluated favorably; a second tier of substandard schools were identified as redeemable only if their deficiencies were corrected. The remaining one-third were singled out and recommended for closure for not providing adequate training. As a result of Flexner's efforts, medical education shifted away from a for-profit enterprise to one that focused on academic study, where physician-researchers in the university setting were responsible for generating new knowledge in the laboratory or the clinic.[2]

The reorganization of medical training towards the tireless pursuit of knowledge has pushed the field forward in many ways. Mechanistic studies have uncovered thousands of biochemical processes necessary to maintain homeostasis. Pharmaceutical medications have been engineered to adjust biochemical processes in the body. Using new technologies, modern medical achievements include replacing ailing joints with artificial ones and repairing small nerves and blood vessels with microsurgery. The human genome has been decoded and researchers are now finding ways to regulate how genetic code is expressed.

Yet while biomedical research is reaching new heights, public faith in the medical field is declining. As the physician's role has shifted to that of a scientist relying on technology that may be expensive, invasive, and even potentially dangerous, the emphasis on healing has diminished.[3] The goal of improving patient care by scientific discovery, either by understanding mechanisms or by clinical trial methodology, has focused our view of medicine towards the accumulation of valuable knowledge, but as a result this type of knowledge has become an end in itself instead of a means to improved care. As complex modern health problems are not easily solved by researching, isolating and changing one variable where multiple systems may be involved, patients with chronic conditions continue to suffer. They become frustrated when treatments with positive research study findings do not provide the relief they expected.

To complement the forward leaps in our growing medical knowledge, the study of individual cases can illuminate how we can apply this knowledge in a patient-centered manner versus a population-centered focus. A case report about an elderly patient with multiple comorbidities can offer a rich description of her condition, empowering a physician to decide which pieces of evidence are most applicable to a similar case. A case report about the patient who does not respond to typical migraine treatment might describe an alternative treatment that could offer help. The study of mechanism and the scientific method are wonderful drivers of scientific knowledge but may fall short as the driver of medical care. Case reports may help temper a blind reliance on this approach and address its shortcomings by reducing the overemphasis on technology to solve medical problems, returning the focus of health care to understanding how to apply evidence to match the needs of the individual patient.

Mechanism, Science and Acupuncture

In biomedicine, researchers studying mechanisms develop experimental medications, create state-of-the-art devices to monitor or improve health, and devise advanced techniques to surgically repair the human body. The media often attempts to make these discoveries sound more exciting or credible by including the catchphrases 'scientists have discovered' or 'research has shown.' The public is told that a new type of medical treatment is the 'best that science has to offer.' Accordingly, readers would logically expect a magazine article entitled the 'science of acupuncture' to spend the bulk of its time explaining the biological mechanisms that occur when an acupuncture needle is inserted. The common use of the word 'science' has reduced the term in many instances to an understanding of mechanism.

Not long ago the idea that inserting acupuncture needles into specific locations on the body to produce healing effects was considered quackery, as no plausible explanations of the mechanism existed at the time. After a prominent article was published in 1971 by a reporter describing his own experience receiving acupuncture,[4] researchers began to investigate possible physiological explanations for the effects of acupuncture. One of the first breakthroughs in understanding acupuncture mechanisms came in the form of an animal study proving that electroacupuncture increases the concentration of endogenous opioids in the brain.[5] In the 1970s, various researchers exploring acupuncture point anatomy found some similarities between acupuncture point locations and myofascial trigger points and motor points.[6] These trigger points — taut bands of muscle tissue associated with pain conditions — can influence the function of internal organs, and this viscero-somatic connection helped to explain how acupuncture treatment may affect internal organ physiology.[7] Investigations into these early conceptualizations of mechanisms raised patient interest in acupuncture as treatment while encouraging researchers to further their inquiries.

During the following decades, additional mechanisms were elucidated. Research revealed that acupuncture modulates the autonomic nervous system by stimulating the release of neurotransmitters such as norepinephrine, epinephrine, and acetylcholine, and by affecting changes in their turnover rate. Stimulating nerve fibers with acupuncture needles results in an axon reflex which triggers the release of various substances, including calcitonin gene-related peptide (CGRP). This in turn causes local blood vessels to dilate thereby increasing local blood flow. Needling acupuncture points can also reduce the transmission of painful stimuli to the brain by depressing the activity of a part of the spinal cord called the dorsal horn. In 1991, functional MRI (fMRI) was invented and came to be widely considered the cutting edge of technology as applied to studying the brain. fMRI enabled researchers to identify which areas of the brain receive more blood flow at the time of an intervention. Using this technology, a group of researchers discovered that acupuncture modulates the activity of the limbic system and subcortical structures in the brain.[8] Findings from these types of research studies helped to explain the observed effects of acupuncture in ways that were acceptable to the biomedical community.

While mechanistic research has done much to open our minds to the idea that inserting acupuncture needles may cause a variety of specific local and nonlocal physiological effects, there is still no grand overarching mechanism to explain the diversity of effects acupuncture has on human physiology. It is interesting to note that breakthroughs in the basic sciences directly preceded the discovery of acupuncture mechanisms in at least two instances. In 1975, enkephalins were discovered as a class of endogenous opioids, which inhibit pain signals in peripheral nerves.[9] The following year, researchers found that acupuncture treatment caused a measurable increase in production of these endogenous opioids, thereby explaining the pain relief observed by patients.[5] The association of these chemicals with acupuncture treatment significantly increased its credibility in the West, despite the fact that for millennia acupuncture had been activating these same pathways for pain relief without the necessity of identifying the biochemical mechanism.

The prescient nature of acupuncture is also evident when considering that the discovery of acupuncture's effects on specific brain regions was dependent on the invention of fMRI technology. Less than a decade after its discovery, researchers began to use fMRI to observe brain activity during acupuncture stimulation. A study revealed a decrease in blood flow to the limbic system of the brain during acupuncture needling, which is a factor associated with a relaxation response.[8] Another study revealed that stimulation of distal acupuncture points associated with vision (LV 3 *Taichong* and GB 37 *Guangming*) results in activation of the visual cortex in the brain.[10] The fact that blood flow was altered in both of these specific areas was taken as proof that traditional acupuncture had real physiological effects, even in the absence of any biological explanation as to how these effects might occur.

It is interesting to note that acupuncture was viewed with more serious interest after biomedical physiology was advanced enough to allow for the discovery of a class of biochemical compounds that play an important role in pain medicine and technologies were invented that allowed closer study of its mechanisms. One might interpret the timing of these discoveries to mean that one reason some disparage acupuncture as pseudoscience is that our biomedical knowledge is still insufficient to explain how it works. It is likely that in the future after new advances in biomedical knowledge new acupuncture mechanisms will come to light, offering further credence to acupuncture from a western scientific viewpoint.

With this in mind, viewing acupuncture's viability as a treatment modality as contingent on western scientific discovery is quite limiting. It seems almost absurd to characterize the practice of acupuncture as antiquated or 'unscientific', when acupuncturists over two thousand years ago now appear to have been ahead of their time, using techniques to stimulate biochemical pathways that were not understood by biomedicine until the late 20th century. Do we really need to wait to accept the observable positive effects of acupuncture that fall outside of currently understood mechanisms because of the limits of biomedicine's current knowledge base? Acupuncture has had effects on blood flow in the brain for thousands of years. Is it only now that acupuncture can be accepted as a valid treatment because the invention of fMRI enables us to observe blood flow in the brain? And when the cutting edge of science suddenly becomes discredited — as has happened with fMRI and the inability of scientists to replicate their findings according to the standards of the scientific method — does that mean the effects experienced by individuals using acupuncture are no longer genuine?[11,12]

What Mechanism Misses in Acupuncture

Regardless of an incomplete scientific explanation, patients have been experiencing benefits associated with acupuncture for centuries. The effectiveness of acupuncture was traditionally achieved through clinical experience, developed and accumulated through observation of many individual cases. Although the vast majority of cases in a practitioner's experience through trial and error may not have been recorded, the collective experience of practitioners has been passed on from generation to generation through the written case record. Despite the perceived bias and low evidence value typically associated with case reports, the use of the recorded experiences of practitioners throughout history ought not to be discounted. Patients should not have to wait to take advantage of the potential of acupuncture until biomedical knowledge and technology become sophisticated enough to understand it.

Our modern sensibilities cause us to eschew bias in scientific thinking, and to that end modern medicine fixates on quantitative variables and objective physical structure. Acupuncture is practiced in a way that emphasizes the qualitative and the functional, therefore running afoul of those expectations. Yet clues gained through qualitative observation can reflect which functions are impaired. Observation of the tongue, pulse, and palpation of the channels can provide information about a patient's condition regardless of whether any abnormal biomedical measurements have been identified or physical structure exists to explain the symptom. Case reports are often used to describe qualitative effects of treatment, leading to a general impression that case reports are not rigorous. Despite this mischaracterization, examining qualitative outcomes can lead to findings that may be difficult to ascertain using solely quantitative measures for testing. One reason why acupuncture might have an important role in treating pain conditions may be because measuring pain is challenging due to its subjectivity. The mental and emotional aspects of health as integral aspects of a patient's state are not best captured in a series of mechanisms, but through qualitative descriptions accessible in the case report. Indeed, when pain is reduced to a mechanistic solution pathway, tragic results can occur as is seen in the current opioid crisis.[13]

Designing treatment to account for unmeasurable but observable factors essential to acupuncture diagnosis is not something that can be captured in a clinical trial. However, it can be targeted through case reports, since these factors may be present in different ways depending on the individual. Being able to consistently detect and make sense of the patterns of qualitative factors is an important part of acupuncture. Observational and cognitive skills different from those employed in biomedicine, are based on experience accumulated from generations of practitioners and taught from master to student. Treating many patients can help to refine one's practice choices, and both the unwritten and the written cases collectively form a body of knowledge that current practitioners can draw upon to predict the most useful treatments for individual patients. It is therefore unfortunate that case reports have been relegated to the bottom of the evidence pyramid in biomedicine. While they may not have epidemiological power, case reports remain one of the best methods to note important observations and describe practitioner thought processes in a manifestation of authentic patient-centered care.

If society limits acupuncture treatment to medical acupuncture and defines acceptable practice techniques as only those acting through scientific mechanisms that we currently understand, suffering patients will be denied effective techniques whose mechanisms are not yet understood. The problem with solely relying on mechanism is that the scope of what is possible is dependent on our current technologies for measurement and on our current understanding of biochemistry. While knowledge in this field is immense and still growing, there is still much that remains a mystery. Even as biomedicine continues to work towards understanding mechanism, the careful observation of outcomes recorded in case reports can offer ways to help patients without waiting for biomechanical explanations to catch up.

Clinical Trials and Acupuncture

Besides the study of mechanism, the clinical trial is another scientific avenue used to evaluate the validity of acupuncture. Careful design of clinical trials can help scientists to reduce or eliminate biased conclusions and recording data and methods meticulously can theoretically enable the replication of results. According to the clinical trial model, in order to determine which acupuncture therapies are efficacious and which conditions respond well to acupuncture treatment, researchers should design experiments according to the scientific method. This includes identifying a problem, framing a hypothesis, and designing an experiment to specifically test that hypothesis. Patients would be recruited and randomized into treatment groups, a prescribed protocol administered, and results collected and analyzed in order to prove or disprove the original hypothesis.

Designing efficacy studies focused on acupuncture is challenging for a number of reasons. The first design challenge concerns the concept of placebo. When designing a basic clinical trial, patients are randomized into one of two groups. The experimental group receives the *verum* (true) therapy, and the control group receives a placebo or inert therapy designed to mimic the true therapy to the extent where the patient will not know whether he or she is receiving the true therapy being tested. In this way, bias can be reduced. Acupuncture research has been plagued by the lack of a true placebo; a patient in a trial can easily find out whether he or she is receiving actual treatment because it is fairly obvious to see if a needle has been inserted or not. Attempts at placebos include pressing with the blunt end of the needle, pressing with objects eliciting a sharp sensation such as toothpicks, acupuncture guide tubes, or specifically designed non-insertion needling devices that create the visual appearance of insertion but are not actually inserted. Some of these substitutes have been validated as sufficiently convincing to patients receiving placebo treatments in clinical studies.[14] However, the act of stimulating the skin at acupuncture points even without insertion can result in a number of physiological responses similar to those resulting from actual insertion. In one Japanese style of acupuncture, needles are placed on the skin but not inserted; the non-insertive 'placebo' applied above approximates traditional treatment but in a different style. Sham acupuncture treatments have also been used in clinical trials, where needles are placed in locations that are nontherapeutic or at depths and stimulation levels that are nontherapeutic. But sham treatments have the potential to affect at least one category of acupuncture mechanisms — a needle placed anywhere will alter the local circulatory or immune function and also trigger local neural pathways for pain relief.[15] Using artificial needles can cause real effects, which renders the placebo useless as such.

A related difficulty with maintaining strict experimental methodology in acupuncture studies is patient blinding. Patients receiving treatment should be unaware of whether they are receiving actual therapy or sham acupuncture as a placebo. Even with physical blindfolding of study patients, although some sham acupuncture treatments feel somewhat realistic to those who have not previously received acupuncture, maintaining blinding is difficult because study patients may talk to each other about their experience, read about acupuncture, or have tried acupuncture outside the confines of the study.[15] Patient blinding is one of the major potential sources of bias in acupuncture randomized controlled trials as determined by a methodological study examining 368 systematic reviews including 4715 randomized clinical trials.[16] It is nearly impossible to create a convincing placebo that is truly inert and insufficient patient blinding may lead to biased conclusions. The validity of any conclusions made from a clinical trial with an intervention tested without a placebo may be compromised through the introduction of potential bias.

A second challenge in acupuncture clinical trial design revolves around the individualized nature of acupuncture treatment. One of the hallmarks of traditional acupuncture is that each patient is a unique individual and the most appropriate treatment should therefore be designed with that specific patient in mind. A set of points that might be therapeutic for one patient with epigastric pain may be completely ineffective for another patient with the same condition. In biomedical care, this idea might seem obvious. For example, two patients reporting epigastric pain should be treated differently if one has a bleeding ulcer and another has more straightforward indigestion. In traditional acupuncture, epigastric pain is also differentiated before appropriate treatment is selected, but the criteria used for differentiation are not the same as biomedical criteria. In addition, the options for treatment (point choices and combinations) are much more varied. Applying the acupuncture to a treatment group in a fixed protocol goes against the nature of traditional acupuncture, and as such, researchers working with protocols are not testing true acupuncture but only an approximation that may not be appropriate for the patient depending on his or her individual signs and symptoms.

Because of difficulties in designing an acupuncture efficacy study, any final conclusions would then be arguable. Placebos in acupuncture introduce their own confounding factors and protocol-based studies do not reflect the flexible approach to treating the individual patient in real-world situations. Even with the best of designs, the overall credibility of scientific studies beyond acupuncture has been a subject of scrutiny in recent years. Published research findings from both mechanistic and clinical trial designs can be refuted by subsequent evidence, with ensuing confusion and disappointment. It is not even clear how much progress biomedicine is making. There is increasing concern that in modern published research, false findings may be common due to small sample sizes, flexible analyses, and publication bias.[17] In a 2013 study, a dozen doctors from around the country examined all 363 articles published in the *New England Journal of Medicine* between 2001 and 2010 that tested current clinical practice. Their results, published in the *Mayo Clinic Proceedings*, found that 146 studies out of 363 proved or strongly suggested that a current standard practice had either no benefit at all or was inferior to the practice it had historically replaced[18] — yet on average it takes ten years for the medical community to stop referencing popular practices after new contradictory studies disproved their benefit.[19] We often fail to recognize that the process of science is ongoing as new findings supplant old ideas. In fact, this process has happened so many times that the term 'medical reversal' has been coined to describe exactly this concept as it applies to health care.[20] Dr. Sydney Burwell, former dean of Harvard Medical School, stated in a commencement address to graduating medical students: "Half of what you are taught as medical students will in ten years have been shown to be wrong. And the trouble is, none of your teachers knows which half."[21]

Modern researchers are developing ideas to circumvent these obstacles, for example by designing pragmatic trials[22] that evaluate the effectiveness of an intervention in real-life routine practice conditions rather than efficacy trials that test a fixed protocol.[iv] Some study protocols are designed with an individualized element,[15] where, for example, a study participant is diagnosed according to TCM principles and the practitioner may choose from a limited set of points corresponding to the TCM diagnosis. This type of study more closely approximates real-world practice, yet some limitations may still exist (e.g., point pools are limited or allowed diagnoses are limited). Another attempt at circumventing questions about authentic practice is the approach known as whole systems research, a term used to describe research of more traditional holistic therapies that expands beyond a simple procedural intervention to include diagnosis and patient-practitioner interactions.[24] Instead of testing an acupuncture group against a placebo group, some trials of this type test an acupuncture group against a second group receiving standard biomedical treatment, and sometimes against a third group that receives both therapies simultaneously. The new designs appear to more accurately reflect how traditional acupuncture is practiced, yet when a wider range of patients are examined, and when treatment is individualized, the outcomes become more difficult to compare and make generalizations from.

The use of clinical trials to accurately evaluate the efficacy of acupuncture may be politically expedient, as results from these trials may lead to making acupuncture available for patients through state regulations and insurance plans. However, because traditional acupuncture is about individualized care, applying a proven protocol in actual clinical practice may not be an ideal treatment depending on the patient. Even if a clinical trial could validate the effectiveness of individualized point selection methods, it would not add to or improve the traditional practice of acupuncture.

While researchers work towards designing more elegant acupuncture trials to test acupuncture's efficacy for hundreds of patients at the same time, clinicians already engage in testing the effectiveness of approaches to solving problems with patients one by one over the thousands of cases

---

iv Efficacy is defined as the successful performance of an intervention under controlled circumstances, whereas effectiveness refers to successful performance under real-world conditions.[23]

in one's career. Each patient is a small-scale opportunity to test a prediction. While making inferences based on uncontrolled studies may certainly be prone to bias, writing case reports that recount the facts, subjecting these reports to scholarly discussion, and communicating the methods and thought processes to other acupuncturists to test and develop further is another valid way of generating knowledge.

The Pitfalls of Scientific Medicine

The struggle continues to try to support acupuncture practice with science, to define acupuncture in biomedical terms, and to test acupuncture for its efficacy using biomedical research designs. Understanding mechanism and carrying out experiments are both firmly ingrained in our culture as being characteristics of science. The process of science in biomedicine can be slow-moving, but ideally a mechanism is discovered, a technology (for example, a medication) is developed, and then the application is tested to verify true benefit. However, the duality of viewing science as mechanism and viewing science as dependent on valid conclusions from clinical trials can sometimes lead to contradictory realities and problematic biomedical interventions.

One type of error caused by this dual conception of science involves medical interventions that appear to have valid mechanisms but are proven ineffective by clinical trials. For example, atenolol, a beta-blocker patented in 1969, was used in millions of patients for decades to treat hypertension. By blocking the effects of epinephrine, these medications would cause the heart to beat more slowly and with less force, thereby lowering blood pressure. However, a 2004 analysis of clinical trials, including eight randomized controlled trials comprising more than 24,000 patients, concluded that despite successfully lowering blood pressure, atenolol did not reduce heart attacks or deaths compared with no treatment; patients who were prescribed atenolol had better blood pressure measurements but their cardiovascular risk did not change.[25] The questionable assumption was that lowering blood pressure would produce a reduction in mortality, but the complexity of the human body and its myriad other mechanisms was not accounted for. Biomedical treatments are often designed to bring lab values back into a normal reference range, but as in this instance, improved lab values may not necessarily translate to better health. As a result, although the use of atenolol might be considered 'scientific' in the sense that it has a plausible mechanism, it is at the same time not considered 'scientific' because clinical trials have not convincingly proved efficacy. Our confidence in mechanism and our strong focus on the importance of measurable lab values can lead us to false conclusions, years of wasted financial resources, and unnecessary side effects. When a new medication is developed, rather than asking 'how does it work' and focusing primarily on mechanism, the more important questions might be those relating to effectiveness: 'does it work?' and 'for whom does it work best?'

Other examples of assumed mechanisms leading to ineffective treatment abound: in 2001 two studies proved the use of antibiotics had no positive effect in the treatment of patients with persistent Lyme disease symptoms.[26] Procedures and surgeries may also be suspect: in 2019 a study from the Stanford University School of Medicine revealed that patients with severe but stable heart disease treated with medications and lifestyle advice alone are no more at risk of a heart attack or death than those who undergo invasive surgical procedures such as stents or bypass surgery.[27] CPR has even been called into question: a 2010 observational study presented the shocking results that there was no significant difference in survival rate for patients who received rescue breathing as part of their CPR compared with patients who received no rescue breathing and only chest compressions.[28]

Consider the issue of hip labrum repair through arthroscopic surgery. Invented in the 1980s, the number of procedures has risen eighteenfold in the last decade. Yet in 2018 a limited study reported that two years after surgery no significant difference existed between the surgery group and the non-surgery group in terms of pain, disability, and perception of improvement.[29] Designing a clinical trial to test the efficacy of surgical interventions is challenging due to the lack of ethical placebo (a true placebo would involve putting a control group under anesthesia, giving a false incision and exposing the patient to significant unnecessary risk). When performing surgical intervention to repair a physical defect, an inference is made that if the structure is restored the function will return and the patient will therefore be better off. In some surgical procedures this inference may be false, but because of our confidence in mechanism and structure, it may be many years before researchers bother to verify if the final outcomes are actually improved.

The confusion regarding the nature of science can also give rise to an opposite type of situation where the clinical trials prove scientific efficacy while the mechanism is unknown. A drug derived from the willow plant, acetylsalicylic acid (commonly known as aspirin), has been used for its analgesic properties since the times of Hippocrates in ancient Greece. It wasn't until 1853 that the chemical structure of acetylsalicylic acid was described, and in 1876 the first rigorous trial of the willow plant's active ingredient proved its ability to reduce fever and joint inflammation. Despite knowledge of its chemical structure, the actual mechanism for the biological effects of aspirin was not discovered until 1971.[30] For many years, researchers did not understand mechanistically how this compound worked despite the medication satisfying a rigorous scientific trial. A number of medications fall under this same dilemma, including metformin, cyclobenzaprine, and lithium, which have been in common use after successful clinical trials for years before some of their mechanisms were discovered.[31-33] As a result, the use of these medications could be considered 'scientific' (having proven efficacy through clinical trial) but at the same time not 'scientific' (without a known mechanism).

Science, Pseudoscience and Acupuncture

The use of the word 'science' can be confusing as it describes two distinct concepts (mechanism and clinical efficacy) that are not interchangeable. Using the same word for these two concepts gives rise to the incorrect assumption that mechanism and clinical efficacy are related. In truth there is an extra logical step between mechanism and clinical efficacy. After a mechanism is discovered a technological intervention may be created, yet before that intervention is accepted it must be tested through the scientific method. However, as described above, some treatments in common use can have known mechanisms but no clinical efficacy, while others can have proven efficacy but no known mechanisms. For this reason, we need to reconsider the very definition of science.

Even if we maintain that one of the two above conditions needs to be true in order for a biomedical intervention to be deemed scientific, some biomedical therapies have neither proven efficacy nor known mechanism. The prescribing of medications for uses that have not been proven by clinical trial (known as off-label prescribing) has become common practice. A 2003 study showed that for the three leading drugs in each of the 15 leading drug classes, off-label use accounted for approximately 21% of prescriptions.[34] Although these medications have not been approved for these off-label indications, physicians prescribe them believing they may provide more benefit than the available approved treatments. While off-label prescribing is not considered ideal due to increased risk from side effects, the FDA website maintains that "healthcare providers generally may prescribe the drug for an unapproved use when they judge that it is medically appropriate for their patient."[35]

Needless to say, this line of argument opens up the possibility that a number of other alternative therapies — even those with ill-understood mechanisms and no scientifically proven benefit — may be applied according to practitioner experience as long as they are safe. In particular, if off-label prescribing being based solely on practitioner experience is acceptable, then traditional styles of acupuncture based on practitioner experience should also be. We therefore need not explain the biomedical mechanisms of how acupuncture works nor prove its scientific efficacy to derive benefit from it.

However, the repercussions of the Flexner report are still evident today. The world of biomedicine is couched in the language of reductionistic mechanisms, and there is an assumption that only therapies that have been proven by science are counted as evidence of a valid therapeutic intervention. An unfortunate double standard exists, and critics of acupuncture have given it the pejorative label of 'pseudoscience.' Yet if biomedical practice requires a mechanism of action and/or successful clinical trials for each intervention, many biomedical interventions could also be potentially labeled as pseudoscience.

The arguments on what qualifies as science and what is relegated to pseudoscience may fall under the domain of philosophy. One philosophical viewpoint of science maintains that an immature scientific field is defined by its lack of a unifying paradigm, consisting merely of a diversity of theories and facts accumulated without an overarching structure or paradigm. This view assumes that over time and with development a mature unified science will emerge with all theories and facts supporting an organized structure. If this viewpoint is true, we might assume that for the practice of acupuncture to achieve scientific standing it will eventually need to be understood completely from a mechanistic viewpoint, and therefore that clinical trial design will improve to the degree that all of the effects of acupuncture could be precisely measurable without argument or bias. For many reasons this is very unlikely, so is acupuncture therefore doomed to remain an immature science?

In this philosophical context, viewing science this way has fallen out of favor, as sciences characterized by different ways of knowing are now viewed as equally useful. For example, the unified discipline of cognitive neuroscience uses fMRI to investigate brain structure in an organized manner, while the less unified discipline of cognitive psychology uses an aggregation of case experience to generate a variety of original overlapping theories to understand mental function. Both disciplines are useful yet they each approach the topic in vastly divergent ways — ways so divergent that one could not supplant the other. In fact, both complement each other contributing knowledge to a hybrid interdisciplinary cognitive neuroscience.[36] The same could be argued about biomedicine and acupuncture. In biomedicine, the biochemistry framework aggregates new facts that are defined within a unified molecular and structural understanding of the world, while traditional acupuncture is based on a plurality of conceptual frameworks favoring an observational and functional understanding of the world. The concepts in biochemistry are mechanistically defined and scientifically supported, whereas the concepts in acupuncture are derived from aggregations of cases where the links between cause and effect are too complex to reduce to compact mechanistic explanations. For both fields, familiarity with their concepts enables practitioners to explain observed phenomena, make predictions, and formulate practical therapeutic options.

Though biomedicine and traditional acupuncture are based on different paradigms, both also have similar goals — to explain and to preserve human health. These paradigms may coexist, each with its own strengths and weaknesses. The philosopher of science Thomas Kuhn famously argued that due to the principle of incommensurability, if competing paradigms are founded on different principles, their merits may not be rationally compared. Three obstacles in particular stand in the way of logical comparison: 1) methodological (where the comparison and measurement methods differ), 2) perceptual/observational (where perceptual experience is dependent on the paradigm), and 3) semantic (that languages of theories from different times and places may not be translatable).[37] Simplistically, biomedicine is a study of cause and effect in human health based on experimental design. Acupuncture is a study of complex interactions within human health based on practitioner experience and many individual cases. While the paradigms of biomedicine and acupuncture can be used in complementary ways to improve patient care, they use different language and different concepts to understand health and disease. Biomedicine as an analytical and deductive way to approach health and illness is simply different from acupuncture's synthetic and inductive approach. Both of these lenses are different, but both are "equally valid and useful."[38]

Conclusion

As two of the defining hallmarks of science, the notions of mechanism and clinical trial are not always considered in all biomedical decisions. As a result, the use of a good number of biomedical therapies, past and present, could be categorized as unscientific. Where some might argue that acupuncture is not scientific due to the lack of an overarching mechanism or because of the inherent difficulties in evaluating efficacy, the basis for those arguments revolve around the interpretation of the definition and scope of science itself. The lay concept of science is fraught with inconsistencies, and the reliance of biomedicine on scientific evidence is not as unswerving as perceived.

The concepts of science, though not perfect, have enabled significant advancement in our ability to understand and fight disease. In contrast, while biomedicine's knowledge base is dependent on the scientific method, acupuncture's knowledge base is dependent on the observation of cases. The paradigm of acupuncture, based on collective practitioner experience, is a practical system by which we can frame our understanding of human health. As the education of new practitioners involves a transfer of experience, the genre of case reports can be consulted as a written record of experience. The collective case literature is the body of research that truly supports the traditional paradigm of acupuncture. Writing new case reports is an important way to develop and refine clinical practice.

CHAPTER 4

Ways of Knowing in Science

ONE OF THE foundational aspects of modern biomedicine is the idea that any medical treatment should be subjected to rigorous testing to verify therapeutic benefit. While Hippocrates is credited with the invention of the scientific method, it was not until the 17th century that this became common practice, when Louis Pasteur published articles describing his studies on germ theory.[1] Pasteur formed a hypothesis, created a control group, carefully observed outcomes, and recorded and publicized his observations. Designing and carrying out experiments is an important way of building scientific knowledge.

However, some hold the unfortunate assumption that the scientific method is the only way to generate a scientific discovery. Observation is another valid way to study the physical and natural world. One might presume that observation should lead to an experiment, which then may have the potential to support a new scientific fact — the assumption being that the observation is merely a feeder for a more sophisticated proof by experiment. But many activities and facts that we consider to be scientific are not based on experiment. Through observation, an astronomer estimates the approximate size of one of the moons of Jupiter. This fact is based on observation; as yet there is no experiment that can be performed to verify the size of that moon. Through observation, an entomologist also discovers a new type of spider in Australia. Perhaps its DNA is sequenced and found to differ from all other spiders previously documented. Sequencing a DNA sample through biotechnology

75

techniques may be a specialized form of observation, but there is no experiment one can do to verify that this spider is a new species.

A specific historical example of scientific knowledge learned through observation was the discovery of pulsars in astronomy in 1967. Observing through a radio telescope, astrophysicist Jocelyn Bell Burnell discovered significant regular pulses of radio waves emanating from a distant galaxy. After finding a second case of the same type of pulsating energy, Burnell inferred that the pulses were flashes of radiation from a collapsed neutron star, and further observations and measurements supported her conclusions. Dr. Burnell arrived at her conclusions not by studying hundreds or even dozens of these pulsars but by carefully analyzing just a few cases, combined with her experience, knowledge, and intuition.[2] Perhaps the most famous example of scientific discovery by observation is Darwin's theory of evolution. It was based on the insights he gained from a series of carefully conducted case studies, contrasting the features of domesticated animals with their wild counterparts, and noting the variations in the form and function of finch beaks.[3] Both Darwin's and Burnell's insights resulted in significant scientific advancement, not through experiment but by the study of individual cases.[2]

In sum, while the scientific method is one part of the discovery process, it is not the whole of it. Quite a bit of research that scientists undertake does not involve the scientific method at all. In the field of science education, students are encouraged to adopt habits of mind that support scientific discovery. Educators work to establish competence in six basic science process skills: observing, measuring, classifying, inferring, predicting, and communicating.[4] While all of these skills are essential to the scientific method, scientific discoveries that do not rely on the scientific method also require these skills. These skills are used not only in the context of discovery, but they are also are crucial in the context of patient care — both in biomedicine and in acupuncture. It is therefore worth considering these skills in the broader context of scientific discovery: what they are, how they influence the natural sciences, how they guide the practice of biomedicine and acupuncture, and how these scientific skills are used in generating knowledge through writing case reports.

Observing

Using the senses to gather information about an object or event is an obvious skill associated with science. Observation of the natural world has led humans to new knowledge in many fields. People have been observing weather patterns since prehistoric times, hoping to gain knowledge applicable to agriculture or public safety. Observing fossils exposed by weathering can offer insight into natural history. Observing the stars can reveal the nature of the Earth's place in the universe. Much of what is accepted as valid scientific knowledge comes from observation.

In biomedicine, observations support much of our accumulated knowledge about the human body. Anatomical observation during cadaver dissection has revealed the locations of muscles, blood vessels, nerves, and organs. Many diagnostic technologies in biomedicine are more refined observational tools that can help extend our senses. The stethoscope is an instrument used to enhance our ability to observe by amplifying the sounds of respiration and heart rhythms. Ultrasound is a technology that enables us to view the orientation and development of the fetus in the uterus. Other tools can enable observation at the microscopic or molecular level: a pathologist observes cells with a microscope to determine whether the cells are cancerous, a geneticist reads the human genome through using DNA sequencing, and an endocrinologist measures the level of hemoglobin A1c in the blood.

In biomedical clinical practice, observations obtained during the examination of a patient can help clarify a plan for treatment, and past observations (especially those which are found to be consistent over many individual cases) support the vast amount of background knowledge required to understand how to address the illness of individual patients. Observations may be objective (e.g., a fracture in the ulna) or they may be subjective (e.g., shortness of breath). Technological aids are often used to enhance and quantify what physicians can sense and to attempt to eliminate potential bias in patient assessment. While eliminating bias is a worthy goal, the use of these observational tools is not infallible in determining cause and effect relationships in patient care. For example, when observed through MRI, a patient with lower back pain is diagnosed with degenerative lumbar discs. Surgical repair may improve the appearance of the structure, but if the pain was not caused by degenerative discs, the pain will likely persist. If biomedical practitioners rely on observation to evaluate the success of treatment, they must remember to observe not only structure, but also how the patient responds, making sure that an intervention designed to address structural (or functional) issues actually improves the patient's quality of life.

In acupuncture practice, observing a patient directly is key to understanding his or her underlying health imbalance.[5] Asking questions of patients allows access to the patient's own observations — these may be subjective (feeling too hot or too cold) or objective (one bowel movement every three days). Observations may be made from the point of view of the practitioner (who observes the patient wincing with pain) or the patient (who describes the feeling of pain as sharp and radiating). Observing the patient involves both structural concerns (visible swelling of a knee joint) and functional concerns (difficulty standing from a seated position). Traditional diagnostic observations are done without the aid of technology; historically, this type of evidence is gathered using the senses directly. Looking, smelling, listening, and palpation findings are noted during a physical exam (including meridian palpation and pulse and tongue diagnosis). These types of observations are combined to create a picture of imbalance whereby a diagnosis can then be made.

Observation is not only used in acupuncture diagnosis, but it is also used to evaluate patient response. The patient with shoulder pain may observe a reduction in pain and an increase in flexibility, the patient with digestive discomfort may notice less gas and bloating, and the patient with insomnia may fall asleep more quickly. After a patient receives acupuncture, observations of a patient's response can be noted to determine its effectiveness, and future treatments of similar patients may be guided by these observations. Observations made by previous generations of practitioners also help to inform diagnosis and treatment of future patients. While the past observations of many experienced practitioners are committed to memory and may be passed down via the oral tradition, written observations, often in the form of case reports, offer more detail and permanence. Records of these observations can help current practitioners compare their cases to observations of the past to determine prognosis and treatment.

Measuring

The act of collecting quantitative data that describes a property of an object or event is a skill essential to the practice of science. For centuries, people have been measuring parameters such as length, area, volume, mass, time, and temperature. By determining a standard sequence of equal units (such as a series of equally spaced marks on a ruler), comparisons can be made between discrete objects or events. Higher-order parameters are combinations of more basic parameters: for example, density is the ratio of mass and volume, and velocity is the ratio of distance and time. BMI is a specific calculation based on height and weight. Measurements can be made in absolutes (e.g., counting the number of tumors in a patient) or in scales in which a reference quantity is standardized (e.g., measuring weight in kilograms, height in centimeters or temperatures in Celsius degrees).[6] Measurement also adds an element of objectivity to science. Using a thermometer can help verify that the freezing and boiling temperatures of water are consistent. Counting plants with dominant or recessive traits in his experiments enabled Mendel to come up with his fundamental properties of genetics.

In biomedicine, measurement is used to inform diagnosis, treatment, and monitor outcomes. Diagnostically, measurements are taken to diagnose conditions or monitor their severity. For example, oxygen saturation in the blood is measured with an oximeter and body temperatures are taken with a thermometer. A complete blood count (CBC) measures red blood cells, white blood cells, hemoglobin, hematocrit, mean corpuscular volume, and platelets. Urinalysis measures a number of parameters, including red blood cells, white blood cells, bacteria, pH, protein, glucose, nitrites, bilirubin, and others. Treatment dosage is measured carefully, as too little of a medication may be ineffective and too much may be fatal. Measurement makes observations less subjective and helps to limit bias, and a consistent method of measuring outcomes is essential to a rigorous clinical trial. However, because biomedical research by its nature leans so heavily on measurable outcomes in order to avoid bias, conditions where objective data are lacking, such as fibromyalgia, may be difficult to investigate.

Measurement is used to evaluate a patient's baseline health condition, to assess patient outcomes post-treatment, and to compare the two. This can be done in the context of research, where patient outcomes are compared to the baselines for large sample sizes to evaluate efficacy, or in the context of patient care, where clinicians comparing measurements before and after an intervention can determine if a particular therapeutic regimen was effective for a single patient. While the improvement of measured parameters can indicate success in a clinical trial or with individual patient treatment, care must be taken to ensure that measurements do not lead us to faulty conclusions. Using lab values as a goal of biomedical practice may be misguided; one caveat of depending on measured criteria is that for some patients, despite improving lab values, they may find that their symptoms persist or worsen. Observing a normalization of estrogen and progesterone levels in a patient indicates there may be a positive change, but this may not necessarily resolve the dysmenorrhea she experiences. Treating just an abnormal lab value may fail to address its underlying cause.

For patients with medically unknown symptoms, where symptoms are present despite having no lab values out of normal range, our overdependence on using lab values to treat can lead to the inability to conceive a useful treatment.[7] At the same time, overreliance on measurement in biomedicine may also unfortunately reduce the significance of important qualitative observations in clinical practice. Evaluative technologies exist in biomedicine that can measure structure (e.g., radiological studies) but what matters is whether a patient can engage in normal activities without significant pain. Subjective data are not absent in biomedical care; for example, observation of gait and sensitivity to palpation is noted in charts of orthopedists and physical therapists. Subjective observation is a real and important part of biomedical clinical practice and is not only unavoidable but may be an important factor in making clinical decisions. For example, one of the reasons why pain management is so challenging is that pain is notoriously difficult to measure consistently, requiring doctors to rely on qualitative descriptions of pain when recommending a course of treatment.

In traditional acupuncture, the idea of measurement is a present but not defining feature. Many meridian acupuncture points are located with the aid of a proportional measurement unit called the *cun*, sometimes translated as 'body inch.' The *cun* measurement is a flexible measure, its absolute length depending on the individual and on the body part being measured. While this measurement is not absolute, it is still valuable in specifying a treatment location. The *cun* measurement is also used to describe the depth of needle insertion, a key to both effectiveness and safety, which can vary depending on the anatomical location of the point. In Chinese herbology, the weight of each herb in a treatment formula is measured to control the correct proportions between the herbs, and also to determine the overall dosage. But overall the skill of measurement has not been as heavily emphasized in acupuncture diagnosis, which prioritizes qualities over numbers. This may partly be due to historical reasons; many technologies used to measure biomedical parameters, such as the blood pressure cuff and pulse oximeter, were invented only within the last hundred years. In traditional diagnosis, pulses are deemed 'fast' or 'slow' without counting beats per minute. In some schools of Japanese meridian therapy, the strengths of the various pulse positions are compared, with the strongest one(s) determining the overall direction of treatment.[8] In Five-Element Acupuncture, diagnosis is made through ascertaining a patient's color, odor, sound, and emotion, none of which are quantifiable.

Not only are diagnostic criteria in acupuncture lacking in measurement parameters, but outcomes also tend towards being qualitative as well. Historically, because many generations ago acupuncturists did not have the technological means to perform chemical urinalysis or to measure spirometry, they had to rely on their visual observation of the clarity of the urine or the auditory observation of the loudness of the cough.

Because acupuncture is concerned with outcomes that directly reflect the patient experience, it eliminates the intermediary step which may contribute to a false inference that an improved lab value is equivalent to patient improvement. Furthermore, certain criteria for diagnosis in acupuncture rely on the patient's qualitative experience including the type of pain sensation, feelings of heat or cold, and more. Arguably, patient outcomes may be confounded by placebo if measurement is omitted completely, but subjective data may include valuable clues to correct diagnosis and effective treatment. Ideally, a combination of both qualitative observation and quantitative measurement may provide a more balanced assessment of a patient's condition. The lack of precision in measurement may be part of the reason why acupuncture case reports have traditionally been less convincing to biomedical audiences and why acupuncture's acceptance in the West has been relatively slow. On the other hand, one reason why acupuncture may be a good complement to a biomedical care model is that the qualitative observations which may be omitted due to biomedical objectivity may offer further insight into finding effective treatments. Case reports are an ideal vehicle to capture these qualitative factors in diagnosis and outcomes, and while they might be missing statistical power, the knowledge gained from reading a case report may nonetheless have the potential to improve care.

Classifying

Classification is an activity that humans have used for centuries to understand the organization of the natural world. Eighteenth-century Swedish naturalist Carolus Linnaeus developed a systematic framework for the classification of organisms. Chemical elements with similar properties are classified into groups in the periodic table. Rocks and minerals are characterized by their origin, grain size, and hardness. Stars are classified according to temperature and mass. Electromagnetic waves are classified according to their wavelength, energy, and frequency. Grouping according to characteristics and noting the similarities and differences between items within the group, helps us to organize and better understand the natural world.

In biomedicine, understanding the classification of bacteria can help researchers determine which types are susceptible to treatment by which antibiotics. Molecules in the body are classified by chemical structure — carbohydrates, proteins, lipids, and nucleic acids. Tissues and organs can be classified by organ system and diseases can also be classified by specialty: gastroenterology, immunology, endocrinology, ophthalmology, etc. With regard to treatment, pharmaceutical companies develop new drugs in classes of chemicals with similar functions. Research studies can identify groups of individuals who might be genetically susceptible to specific conditions such as certain types of breast cancer.

In biomedical clinical practice, classification is performed every time a diagnosis is made. When making a diagnosis, a physician places the patient in a category according to signs and symptoms. One benefit of categorizing in this way is to help the practitioner to recall and apply effective treatment. With the seemingly exponential growth of biomedical knowledge, resources for physicians (e.g., the 2000 page reference text Current Medical Diagnosis and Treatment[9] and online resources such as uptodate.com) can help them to identify the disease category a patient falls into as well as listing treatment options in a condensed fashion, thereby not requiring a time-consuming literature review with every patient contact. In most cases, classification increases efficiency in patient care through fast and accurate referrals.

However, classification can occasionally hinder patient care. A patient with wrist and hand pain may be incorrectly categorized as having carpal tunnel syndrome and be referred to a hand surgeon. She may then receive surgery but fail to obtain relief because her cervical spine (an area outside the specific expertise of the hand surgeon) was the origin of pain. Some of the most challenging disorders to diagnose, such as metabolic disease, involve more than one system. Hypertension in a patient may certainly be relevant to the heart but also the kidneys or lungs. Because knowledge has been compartmentalized, it may be more difficult to find solutions to complex multifaceted health problems that defy simple categorization.

Classification is a skill frequently used in acupuncture practice as well. Acupuncture points are classified by meridian (e.g., 67 points on the

Bladder channel, 11 points on the Lung channel), with numerous other point categories existing across the regions of the body, including *xi*-cleft points, *mu* alarm points, *jing*-well points, fire points, lower *he*-sea points, and back *shu* points. The groups that each point is a part of indicate to some extent its usefulness in specific situations. *Zang-fu* organs are grouped according to five phases — metal, water, wood, fire, and earth — with the phase of a given organ reflecting its characteristics and its relationships with others. Diagnosis in TCM style acupuncture is an act of classification — the grouping of signs and symptoms in order to find out what organ is most out of balance, and in what way (deficient/excess, cold/hot, yin/yang, interior/exterior).[10] Beyond acupuncture, individual herbs and herbal formulas are categorized according to function (e.g., moving Blood stasis or extinguishing Wind). Without classification, the traditional practice of acupuncture would be a disorganized smattering of folk remedies. Classification systems aid in recall and in proper application.

Classification skills are less important in the context of outcomes in acupuncture compared to diagnosis and treatment (aside from the idea that if the symptoms are not classified correctly good outcomes are unlikely). But case reports can be an important way to help clinicians improve their classification skills in the context of patient care. Parsing through details of a difficult diagnosis may help guide an acupuncturist to understand into which disease category the symptoms fit or which category of points to use. Noticing the differences between cases can help refine classification skills, while grouping similar cases identify overarching themes. While the accumulation of individual cases cannot by nature prove any universal principles, the analysis of commonalities and differences in a set of cases can form patterns that may be useful in improving patient care.[11]

Inferring

Inferring is the process of generating an explanation for an event or process, based on previously gathered data or information. Scientists can add to human knowledge of nature through making inferences. By comparing the bony structure of fossil dinosaurs to that of modern vertebrates, paleontologists can make inferences about how dinosaurs moved. Even though electricity cannot be seen, inferences can be made about the flow of current from watching bulbs lighting up when connected to a battery.[12] Scientific knowledge in the field of geophysics is generally acquired without the benefit of controlled experiments, relying instead on the timing of proposed cause and effect phenomena to infer causality.[13] The process of inferring often involves creating a model — a physical, mathematical, or conceptual representation of a system.[14] Physical models can be used to visualize a structure which is difficult to observe directly, such as the nanoscopic model of the atom. Mathematical models involve a set of equations that can help to explain how the natural world behaves, like Newton's second law (usually written F=ma). And observations of light phenomena gave rise to two different conceptual models: the physical (particle model) and energetic (wave model).

Many inferences made in biomedicine ultimately led to the creation of models. In the 1700s the importance of eating fresh fruits and vegetables to prevent scurvy was based on inference, where it was noted that people who ate citrus fruits were not affected by the disease. Despite the fact that the chemical structure of vitamin C was not fully determined until 1933, using diet to prevent scurvy was a cure determined by inference.[15] The inference made regarding the connection between lack of citrus and scurvy led to a nutritional model, where eventually adequate intake of vitamins and minerals was recommended to maintain healthy function. The pharmacology model is used to identify the molecular composition of biologically active compounds in order to understand its effects on living systems.[16] Without the use of this model, while physicians could certainly observe the effects of ingesting certain natural substances, it would not be possible to create new synthetic life-saving drugs or produce them in mass quantities.

Yet while models are based on factual premises, no model can be completely accurate, and models are always subject to revision when individual situations arise where the explanatory model does not match observations in real life.[17] For example, physicist Linus Pauling originally proposed an incorrect model of DNA with three strands, only to be supplanted by the more accurate and now accepted two-strand model explained by Watson and Crick.[18] Established human cancer cell lines are routinely used as experimental models in cancer research, but while observation of cancer cells *in vitro* can provide important knowledge, there are limitations to these models when applying this knowledge to *in vivo* tumors.[19] Animal models are used in virtually all fields of biomedical research including basic biology, immunology and infectious disease, oncology, and behavior,[20] yet they too have limitations, with less than 1/10 of the conclusions in animal models applying to humans.[21]

Many of the inferences made from models have led to the development of useful therapies. Diagnoses in biomedicine are made in the context of specific knowledge about the relevant structures of the physical body within the cell/tissue/organ model. For example, if a patient exhibits sudden onset weakness on the left side of the body, slurred speech, and a history of hypertension, an inference could be made that the patient may be suffering a stroke. According to the model of the brain which maps function to location, further testing would ensue, including a brain scan that looks specifically at the motor regions on the right side of the brain, as well as regions related to speech.

Yet false inferences in biomedical clinical practice in some instances may lead to ineffective treatment, the use of non-beneficial interventions, and even iatrogenesis. When a new therapy brings patients' unhealthy lab values back into a normal reference range, researchers infer an increase in health; however, one type of incorrect inference involves assuming that medicine can cause the body to benefit theoretically by means of a logical biochemical reaction. In 1999, the medication Vioxx (rofecoxib) was approved by the FDA and prescribed to millions of patients for osteoarthritis pain. The drug was assumed to benefit the heart because of its anti-inflammatory effects, but in 2004 (after sales of $2.5 billion), it was voluntarily withdrawn from the market by the pharmaceutical manufacturer because it was found to have the opposite side effect — of increasing the risk for heart disease.[22] Errors in biomedical inference can be due not only to our assumptions about normalizing lab values and repairing visible anatomical structures but can be a result of our faith in biochemical mechanism models.

Over time, acupuncturists have inferred knowledge by observing many cases and codifying the knowledge into practical models. Observations of many cases over many individual practitioners help to strengthen useful inferences and weed out or refine inconsistent ones. The types of inferences made in traditional acupuncture theory were often based upon functional rather than structural relationships. The model which the acupuncturist employs to describe these functional relationships is the *qi* model. *Qi* flows throughout the body, and specific imbalances in *qi* are can be used to explain the various states of illness. Observations of signs and symptoms in a patient can indicate to the observant acupuncturist where the *qi* is excess, deficient, or stagnant, thereby providing the basis for designing treatment. Another model important to explaining the relationships between organs is the five-phase model. Described in the *Nan Jing* (Classic of Difficulties),[23] the five-phase model attributes an elemental quality to each of the key organs. Functional weakness in one organ might over time spread to weakness in another via the generation cycle. Sometimes, depending on the organs being observed, weakness in one organ might cause another to overreact and overcompensate via the control cycle. This model helps to explain specific phenomena, such as why an individual patient with a high-stress level and propensity towards anger might have difficulty with digestion (Liver overacting on Spleen).

The most obvious and foremost model used in traditional acupuncture is the meridian map. Over time, acupuncturists noticed that stimulating certain points on the body helped resolve specific symptoms, and points that had actions on similar symptoms and conditions tended to follow a trajectory. Needling meridian points will affect the functioning of related regions along that trajectory. For example, the point LI 4 (*Hegu*) has an effect on elbow pain, shoulder pain, jaw pain, and sinus congestion because the Large Intestine meridian traverses those regions, while a map of the extraordinary meridians explains why GB 41 (*Zulinqi*) can have a beneficial effect on the waist.[24] These maps have been built on accumulated knowledge from generations of practitioners who have generated these models and passed down their knowledge over time.

In the clinical setting, an effective acupuncturist must apply traditional knowledge to make inferences beyond explaining why treatment points have specific effects, into understanding the nature and detecting the cause of disease. For example, seeing the same observations repeatedly — like the presence of a dark vein at the auricular lumbar area on a patient — can be indicative of Blood stasis and pain in the lower back, when a holographic model is applied. Case reports, as a form of descriptive research, can be used to identify explanatory patterns in clinical practice. While they can also be used to generate hypotheses for future research, case reports can model the process of making inferences that enable effective diagnosis and treatment.[25] It may be difficult for biomedicine and the West to accept a non-physical model because without a material basis this type of model is not objectively verifiable. However, if a model is based on and refined through observation of many cases, these models retain their value if they can hold up in the world of prediction.

Predicting

Using information and models to predict future events is another skill essential to the scientific process. By knowing how things have worked in the past, a scientist can project future outcomes. Some scientific predictions are testable; in chemistry, 4 g of hydrogen gas reacting with 32 g of oxygen gas produces 36 g of water, while giving off a predictable amount of energy. This reaction can be reproduced under controlled conditions in a lab. The molecular model can be used to not only explain the reaction of hydrogen and oxygen gases, but it can also help us to accurately predict what happens when we start with different amounts of the reagents. This being said, some types of predictions made in the natural sciences are not testable hypotheses. In these instances mathematical models and computer models take data from the past and enable us to generate predictions. For example, as world temperatures rise due to global warming, ecologists may predict more wildfires, drought, and an increase in the number, duration, and intensity of tropical storms, along with a rise in sea level. The graphical model of global warming is an observable trend, but there is no ethical experiment that could verify the effects of global warming (no experiment can purposely raise the earth's temperature 2 degrees Celsius worldwide to be able to prove that the calculation of predicted sea level rise is correct). Complex models in modern science can include many variables and computer models are increasingly being used to model environmental and ecological systems. By being able to integrate many variables, use information from a variety of sources, and account for missing data and uncertainty, forecasting of future events may be more reliable.[26]

Predictions in biomedicine can be simple or complex. For example, a pattern-recognition algorithm may help to predict which individuals may be diabetic. If an individual presents with frequent urination, increased thirst, neuropathic pain in the extremities, or retinal damage, the presence of several of these signs and symptoms in concert leads the physician to explore further diagnostics. With diabetes, researchers observing patterns in the regular fluctuations of blood glucose over the course of a daily rhythm may be able to develop dosage timing guidelines or even technology in the form of an insulin pump, through predictions based on previous data. With computer-assisted data collection and analysis of multiple parameters on many patients, subtle relationships may be distilled to make predictions identifying the future trajectory of the disease. Analysis of retinal scans of 300,000 patients allows for a more specific prediction of cardiovascular risk and blood pressure. Dermatologists can better predict if a patient's skin lesion is malignant through a comparison of over 130,000 pictures previously analyzed by computer software.[27] Even electronic health record (EHR) systems are making data more available to researchers for analysis and to support decision-making and guide policy.[28] However, while it is easiest to formulate a prediction by collecting data isolating cause and effect relationships by changing one variable (in the style of the scientific method), with many conditions more than one variable can be at play. A single variable may turn out to be insignificant by itself but may become significant when juxtaposed with other variables.[28]

The predictions associated with biomedical patient care have two general categories: treatment predictions and prognosis predictions. Both types of predictions are common to clinical practice in biomedicine. Any therapeutic choice by a biomedical physician involves a treatment prediction that considers which intervention is likely to offer the best benefit to the patient. A physician may choose one therapeutic procedure over another based on which therapy has better outcomes in clinical trials or based on whether or not the patient is more likely to experience side effects. Matching a patient to a treatment is based both on information from the past and knowledge of the present circumstances. Screening tools or biomarkers may also help a physician determine which patients will be better candidates for certain procedures and medications. Hundreds of thousands of biomedical articles have been published to describe what we understand about the effectiveness of therapies; in the clinic, the success of a treatment prediction will rely on the physician's knowledge of the literature, combined with their experience and evaluation of the patient at hand.

The second type of prediction associated with clinical practice is that of prognosis. An experienced biomedical physician can predict that the condition of an individual patient may be likely to improve or decline. The patient may inquire as to the success rate and recovery period of a procedure. Depending on the data, these predictions may vary in precision (whether or not there is a range of likely outcomes) and accuracy (how correct the prediction turns out to be). Prognosis predictions may be based on outcomes or demographic data; depending on the biomedical specialty and condition, accurate prognosis may be a difficult task. In some fields, the prognosis is so difficult to predict that researchers studying prognosis have gone so far as to recommend discussing potential scenarios rather than offering a prognosis range.[29]

In acupuncture, many of the models used for treatment design are predictive. Understanding multiple models can give a traditional acupuncturist options for choosing effective treatment points and combinations depending on the case. For example, for a patient with fever, loud cough, and rapid pulse (Lung Heat pattern), a TCM model could be used to predict that needling a point near the distal end of the Lung channel (LU 11 *Shaoshang* or LU 10 *Yuji*) may bring relief. Within a

Five-Element treatment model, another patient with a greenish hue, a loud and shouting voice, and whose predominant emotion is anger would likely respond to treatment focusing on the Wood element. The models that these predictions are based upon were developed and tested by generations of acupuncturists and based on actual clinical cases, both written and unwritten. Innovative thinkers can adapt the theory to expand the predictive power of the model. Yoshio Manaka, a 20th-century Japanese practitioner, suggested using some non-standard locations for front *mu* diagnostic points to predict which organs would be most useful to address based on his experience seeing thousands of cases.[8] Kiiko Matsumoto has similarly generated new ideas on using classical concepts to predict which diagnostic zones might be affected in patients with thyroid and adrenal conditions and which points may help relieve them, again based on her extensive clinical experience.

The case report can describe specific considerations when predicting the effectiveness of certain treatment goals or acupoints. Because of the variety of styles and the dearth of conversation as to which acupuncture approaches may be more useful for a given patient, compiling and comparing case reports that describe the use of different styles may help future clinicians navigate the diverging currents of modern practice. Analysis of case reports may reveal which applications are effective, and which ones may be discarded due to poor outcomes. Acupuncturists compare outcome predictions when choosing which style or which point prescription may be most likely to elicit the quickest recovery. Determining the best course of treatment, how many sessions per week, the duration of needle retention, and whether or not to use electroacupuncture are all treatment specifications chosen based on experience. Every single case in clinic is an opportunity to test a prediction, and gauging the success of those predictions through observing outcomes over time refines practice.

Given that health providers are not infallible, some predictions in acupuncture end up being incorrect; prognosis is often a difficult task especially for novice acupuncturists. But as practitioners gain more experience, their ability to make good predictions using theoretical models does increase and leads to more effective treatment. Also, by pooling resources through aggregating case reports, acupuncturists may become more precise and accurate in their predictions. Reading a variety of case reports can extend each individual practitioner's experience and enhance a practitioner's ability to make both treatment and prognostic predictions.

Communicating

The use of words or graphic symbols to describe an action, object, or event, is an important skill that enables us to generate and transmit accumulated knowledge. Communication between scientists and the public includes sharing knowledge of scientific content, familiarity with the scientific method as a process, and awareness of the impact of science on society. Individuals besides scientists involved in the dissemination of scientific information include journalists and educators, decisionmakers in government and scientific institutions, and of course the general public.[30] Scholarly communication, among both academics and professionals promotes the exchange of ideas in the context of professional journals and other publications, libraries, educational institutions, professional organizations and events, and electronically via the internet.[31]

In biomedicine, communication of new ideas is accomplished primarily through the avenue of publication. Over 30,000 peer-reviewed journals are cataloged on PubMed, with two million papers published per year. Sharing results of clinical and mechanistic research is key for the development of the medical knowledge base. Biomedical texts are revised and updated frequently; the internet has enabled physicians to continually update the corpus of biomedical knowledge through online clinical resources as well as through hundreds of online peer-reviewed publications. Communication in the biomedical field also occurs through oral transmission, as medical students learn the critical thinking skills that are necessary to apply the knowledge base that exists. Conventions allow for the dissemination of new knowledge, and seminars and continuing education courses allow practitioners to learn new procedures and manual skills. In practice, clear communication among multiple collaborating providers can lead to more comprehensive patient care, as well as an enriched learning experience that can allow us to provide better care for similar patients in the future.

In the early development of traditional acupuncture, knowledge and skills were transferred via the oral tradition. Communication between master and apprentice allowed for the younger generation to learn how to interpret written theoretical texts and apply the critical thinking and manual skills necessary to deliver successful treatment. One advantage of oral communication is that with the transmission of information, the student can ask the master about any unclear teachings or texts. Indeed, while the older Chinese medicine writings are conveyed in a manner that allows for interpretation and flexibility, they are also somewhat ambiguous and require guidance in their application. This mode of transmission becomes particularly crucial if the original language texts are not in one's native language; the connotations of the English words used to represent Chinese concepts may lead to misconceptions or interpretations untrue to the source text. Even for readers fluent in Chinese, classical medical texts use grammatical forms that are not in common use today and are subject to subconscious assumptions about Chinese medicine by layering on a biomedical meaning.

Because the philosophies underlying the models are also different from standard biomedical concepts, understanding how to apply traditional theory can be aided by thoughtful case reports where theory is applied successfully. Writing and examining case reports that compare and contrast treatment approaches may lead to common ground between different factions, giving rise to a more unified field when considering educational policies and research design. Through written communication, audiences can be reached over great distances, even many years later, but misunderstandings may result if the writer's thoughts are not clearly expressed. For this reason, case reports must be written as clearly and thoroughly as possible, as it is difficult for the reader to ask questions for clarification, especially many years hence.

In the modern world with its international community of acupuncturists, writing and reading about how to apply these concepts successfully in real-life situations can perpetuate and deepen the practice of acupuncture.

Modern theoretical texts abound in numerous languages, and acupuncture journals discuss both theoretical ideas and their applications in case reports. Communication in the form of a case report is a record that can be written to explain the application of critical thinking skills. Publishing a complete and rigorous case report communicates it properly to the field, making it accessible to all for posterity, and ensuring that the reader applying the information does it correctly. Furthermore, case reports may communicate ideas with broader implications, which may raise discussions regarding healthcare delivery, ethics, safety, or policy.

Conclusion

Some might argue that the scientific skills that have been used to accumulate knowledge for millennia are all employed in the current scientific methodology that is required for designing and carrying out a scientific experiment. A mechanistic study would not be possible without observing, measuring, inferring, or predicting. A clinical trial would require the classification of test subjects according to inclusion/exclusion criteria, creating a hypothesis as a specialized type of prediction. Key aspects of good experimental design, including control of variables and randomization, enable practitioners to use the conclusions of clinical research studies to feel more confident in making inferences and predictions in practice. One might assume that since experimentation requires all of these skills, it is the ultimate pinnacle of achievement in science. However, we would do well to remember that there is quite a bit of scientific knowledge that was never acquired through experiment. This knowledge should not be discredited because by its very nature it doesn't conform to the requirements for evidence that have been in favor for only the last hundred years. While the logic of the scientific method can help us to reduce significant bias in the discovery of new ideas, the scientific method is not infallible. Certainly, clinical trials will continue to be performed, and the findings from this method will certainly add to our understanding of human health and increase the quality of health care. But considering a new understanding of the nature of science should not threaten those who work to understand human health through the scientific method.

The noted 20th-century astronomer Carl Sagan once remarked that "Science is a way of thinking more than it is a body of knowledge."[23] Observing, measuring, classifying, inferring, predicting, and communicating are the basic skills of science that inform this way of thinking. Acupuncture is based on thousands of years of experiential evidence aligned with these skills for thinking, and all six science skills are used when writing a case report as well. It is necessary to observe the patient carefully to be able to describe him or her in detail. A thorough practitioner measures certain parameters before treatment and compares them to those taken afterwards to verify the therapeutic result. The act of diagnosis requires the ability to classify and differentiate in order to understand the pattern of imbalance. Inferences may be made about the etiology and pathogenesis. The choice of treatment involves prediction and making choices in ways that preserve individuality and patient-centeredness. And finally, writing case reports is a way to communicate newfound knowledge with others. As all six skills are involved in acupuncture case reports, they should be considered a rich illustration of science as a human activity.

CHAPTER 5

Case Reports and Integrative Medicine

TRUE INTEGRATION OF traditional acupuncture with biomedicine is a multifaceted process, where theoretical, practical, educational, and political considerations all contribute to the manner in which acupuncture is practiced. Acupuncturists and biomedical physicians have worked alongside each other for many years, and over time have developed ways to collaborate. A number of reasons have contributed to the trend towards integrative practice over the last several decades. As medicine has increased in complexity, there has been growing recognition that teams of providers practicing a variety of modalities may provide more comprehensive care for patients. At the same time, patients — armed with information from the internet — are seeking alternative options as they begin to realize the limitations associated with solely pursuing biomedical treatments. Both patients and providers are gravitating towards less invasive and less risky treatment options, particularly given the high numbers of adverse effects from pharmaceutical interventions and complications associated with surgery. Many have found that complementary and alternative medicine (CAM) therapies may offer additional treatment options for diseases that do not respond well to biomedical therapies.[1] Because of the potential benefits of including CAM therapies in conventional practice, many researchers, educators, and clinicians have been working in their respective areas to support the practice of integrative medicine.

True integration of acupuncture and biomedicine faces a number of obstacles, due to the fact that traditional practice of acupuncture requires a perspective on health that is radically different from the reductionistic biomedical construct. Students who attempt to theoretically integrate the two may inadvertently conflate biomedical concepts with traditional ones, potentially leading to ineffective treatment. For example, the suffix '-itis' indicates inflammation, which implies a Heat pattern (because of the Latin root '-flam', meaning heat). However, a number of inflammatory conditions are no longer diagnosed by the sensation of heat; for example, osteoarthritis is typically diagnosed by x-ray or MRI. Despite its being categorized as an 'inflammatory' condition, many patients with osteoarthritis often experience more pain during cold weather, and as such would be diagnosed with a Cold condition. If an osteoarthritis patient is misdiagnosed with a Heat imbalance, a practitioner might use inappropriate acupuncture points and techniques, likely missing out on the healing potential of moxibustion. Integrating acupuncture with biomedicine by redefining traditional theory as biomedical concepts may lead to an inauthentic version of acupuncture that dilutes its strength.

Beyond distorting traditional practice, using integrative language to communicate with the public as well as with other practitioners misrepresents the capabilities of acupuncture therapy. In practice, this is all the more evident since the general public comes to acupuncture with a western bent. Many practitioners employ biomedical terms such as 'increasing blood flow to local areas' or 'boosting the immune system' when explaining the effects of acupuncture to patients, without recognizing important guiding principles such as the interactivity between organ functions or diagnosis by pattern. Similarly, when acupuncturists and biomedical physicians discuss acupuncture in patient care, the language is often couched in biomedical terminology but with use of more specific mechanisms. This type of translation between paradigms may result in an interpretation of acupuncture that leads away from authentic practice.

Assimilation or Integration?

Above and beyond the challenges of merging two very different philosophies of care is the question of what constitutes integration. One of the most prominent textbooks on integrative medicine describes it as an approach that "takes account of the whole person (body, mind, and spirit) including all aspects of lifestyle" and noted that this integrative approach "makes use of all appropriate therapies, both conventional and alternative."[2] This approach to integrative medicine is more akin to assimilation instead of integration. Assimilation, a process where one entity is absorbed into another, can involve the loss or sacrifice of aspects of one's identity, often for the purpose of reaping benefit from participation in a predominant system. Integration, the process of combining one entity with another to form a whole, brings groups into equal participation, respecting the essence of each side and working to benefit all participants. Positioning acupuncture as the alternative implies that it is not on par with biomedicine. This is undoubtedly due to conventional medicine being the more authoritative player culturally and politically, which unfortunately dictates the structure and framework into which CAM therapies are being assimilated. Furthermore, beyond medical acupuncture texts that attempt to expunge traditional theory from acupuncture practice, a small but growing number of publications attempt to roughly translate traditional acupuncture ideas into biomedical terms.

While the exploration of the connections between biology and traditional acupuncture theory is certainly fascinating, multiple caveats arise when integrating the two disciplines by defining acupuncture in biomedical terms. The practice of traditional acupuncture is based on observations of patients as well as the theory built through countless individual predictions about patient illness either confirmed or refuted by observed outcomes. In the contrasting world of medical acupuncture, acupuncture points are chosen because of their mechanisms or because specific protocols have been proven statistically efficacious in clinical trial. While the medical acupuncture approach may be helpful for some patients, the focus on observation is lost, and its fixed protocols do not allow for flexible application based on observation. In addition, defining the objective of in acupuncture treatment as a biomedical action may result in overlooking other concurrent mechanisms that may be contributing to effectiveness. For example, acupuncture treatment may reduce pain through enkephalin modulation, but other mechanisms are also at work. And finally, aspects of traditional theory without a potential biomedical analogue may be viewed with suspicion or downplayed despite potential effectiveness. Traditional theory, which might be difficult to rationalize within the biomedical perspective, may be seen as lacking credibility and therefore is discounted in practice, where the reason why a relevant bio-medical mechanism might not yet exist is because it has not yet been uncovered in research.

The liabilities of the assimilationist approach can be seen quite clearly concerning the education of biomedical physicians. In 2015, a survey of U.S. medical schools revealed that approximately half offered at least one CAM course or clerkship but only 11% included inter-professional education with CAM providers.[3] In a survey of pain medicine fellow-ship directors, only half reported that their fellows had even observed acupuncture.[1] In the aforementioned integrative medicine textbook of over 1100 pages only seven are devoted to acupuncture; of nearly 150 con-tributors, not one licensed acupuncturist is listed. Its table of contents lists acupuncture under "Tools for your practice" in the biomechanical category, with short chapters describing acupuncture's use for only two conditions — headache and nausea — despite the growing literature of clinical studies for many other conditions.[2] Needless to say, omitting the traditional basis for acupuncture practice denies centuries of collective experience and removes its patient-centered focus from consideration. As a result, many of the strengths of acupuncture are being eliminated rather than integrated.

It is true that engaging in mechanistic research is interesting in that it may give us insights on how to improve acupuncture practice. For exam-ple, noting the blood chemistry after applying certain frequencies of elec-troacupuncture can enable us to more precisely target the desired effects. However, explanations such as 'acupuncture relieves pain by increasing the body's production of endogenous opioids' eclipses traditional knowledge with biomedical mechanism, thereby assimilating acupuncture into a biomedical construct. In the rush to adopt mechanistic explanations, the traditional theoretical understanding of how an acupuncturist creates an effective treatment falls by the wayside. If acupuncture treatment is delivered as protocols by health care providers without training in traditional approaches, the ability to flexibly apply the fruits of thousands of years of collective experience to predict useful treatments for new situations is lost.

The difficulties of the assimilationist model do not end at the theoretical and practical level but can even extend to the modern biomedical system of referrals and clinical care. The chief advantage of the referral system is that each practitioner provides treatments in his or her focused area(s) of expertise. Individual practitioners thus would not need to spend extra time and energy to become effective practitioners of a wide range of modalities, and patients would receive care from properly trained providers who ostensibly get better results and make fewer mistakes. At the same time, the quality and quantity of referrals depend on a variety of uncertainties including practitioner and gatekeeper bias, communication and networking skills, and familiarity of each practitioner with other CAM therapies. Individual patient outcomes may depend on the style of acupuncture used or the effectiveness of the practitioners administering the acupuncture. Obtaining useful EHR data from patients receiving complementary care help may also seem elusive due to the wide variety of patient presentation, the diversity of styles, and practitioner experience.

The promise of the benefits of any integrative model may be hampered by issues of coordination, including poor communication, lack of clarity around roles and responsibilities, disagreements regarding approaches to patient care, and interpersonal dynamics.[4] A 2017 survey by the National Commission for the Certification of Acupuncture and Oriental Medicine (NCCAOM) indicated that the majority of Chinese medicine practitioners expressed a high level of interest in working in integrative medicine, but major concerns were raised regarding the influence of governance and hierarchy of each individual center and its effect on patient access to acupuncture.[5] For example, a 2014 survey of pain management fellowship directors reported that two out of 65 directors cited "insufficient evidence base for acupuncture," indicating that despite general acceptance by pain management directors, particular centers may suffer from lack of acupuncture offerings simply due to the personal opinions of gatekeepers.[6]

The biases of a clinic owner may also radically influence the patient's experience of acupuncture, for example, if medical acupuncturists are hired preferentially over traditionally trained licensed acupuncturists. This concern also applies to hospital hiring, as many hospitals have rules regarding consideration of hospital privileges, making it potentially easier for MDs, RNs or physical therapists to be hired to do acupuncture procedures without having received full acupuncture training. An uncoordinated team may lead to increased costs of overlapping treatments, and potential iatrogenic effects such as herb-drug interactions may occur.[6]

It may be politically and culturally unrealistic to consider the traditional acupuncture paradigm as having parity with the biomedical paradigm. However, if acupuncture is assimilated into instead of being integrated with biomedicine, the underlying theoretical basis of acupuncture may be stripped away from practice, thus sacrificing the strengths of acupuncture as a healing modality. It is important to address inadvertent transformation and consequent loss of the strengths of the Chinese medicine paradigm due to assimilation. Defining acupuncture by mechanism and prioritizing the biomechanics of acupuncture may result in the theory slowly being lost and relegated to a curiosity. The chain of experience transmission will be broken as traditional acupuncturists lose thousands of years of accumulated experiential knowledge. The paradigm of traditional acupuncture, the theory which supports its patient-centered practice, is in danger of being lost in the transition to integrative medicine. Although a variety of integrative practice models exist under the pretense of making the best of both worlds available, some of these models are developed with referral and business relationships in mind rather than by utilizing of the strengths of each modality with the ultimate goal of improving patient care. If it were possible to remove politics, financial constraints, and other obstacles that seem to interfere with doing good medicine, what might integrative medicine look like?

Practical Integration through Case Reports

It makes sense to begin a discussion of the potential of integrative medicine by recognizing the strengths of each paradigm rather than attempting to assimilate acupuncture piecemeal into a dominant delivery system. In this way, integration could be more systematic and systemic, giving rise to a more comprehensive vision of integrative medicine where "both bridgeheads stand solidly grounded in their own foundation."[7] Envisioning and contemplating what both systems bring to good patient care is the right way to implement an integrative practice arrangement. Rather than forcing acupuncture into the methodology of experiment and relying on potentially inconsistent referrals, writing and reading case reports from an integrative stance may offer a more practical way to collaborate.

Writing integrative case reports about individual patients and their experiences receiving combined care offers a practice-based approach to help improve the quality of integrative medicine. To truly integrate medicine means that alongside the biomedical approach to health care, acupuncture is performed according to its own paradigm and on its own merits. Genuine collaboration between practitioners therefore requires a case-by-case consideration, keeping in mind the primary goal of improving patient care. Specific case reports may demonstrate when referrals may be warranted and when effective practical application of combined therapy is truly patient-centered. This approach gives practitioners the flexibility to apply acupuncture specifically tailored to the individual case. Preserving the integrity of traditional acupuncture and bringing its strengths to the table thus requires careful consideration of the way in which integrative case reports are conceived.

In biomedicine, case reports are written for a variety of purposes, including alerting other providers to unusual presentations or offering a new treatment that may potentially be effective. Similar purposes exist for writing acupuncture case reports, but in the case of integrative medicine, it may be useful to also consider writing acupuncture case reports for a biomedical audience or an audience of complementary providers. Case reports may help to educate professionals in other disciplines as well as educate other acupuncturists about the potential for effective treatment by presenting an actual example. For biomedical publications, this does not mean defining acupuncture and the treatment merely in mechanistic terms, but rather writing in a way that respects the thought process and educates on the importance of the traditional approach. This can then allow for a biomedical practitioner to view an acupuncturist as a partner rather than the acupuncture treatment as an intervention.

Writing a case report in this way may also enable the integrative physician to know when to refer. In an albeit limited survey of acupuncturists in an integrative medicine clinic, those queried indicated that even if physicians were willing to refer, they usually knew little about what conditions acupuncture could be used for and therefore when to refer.[8] Although these types of case reports might not be intrinsically 'integrative,' as they are written mainly about the potential benefits of only one modality, sharing examples from these types of case reports may be useful to enhance the coordination of care.

Integrative case reports demonstrate how acupuncture interventions can be combined collaboratively with treatments from other disciplines to enhance the possibilities attainable through working together. Case reports demonstrating improved outcomes after applying two modalities concurrently or in sequence not only address assimilationist attitudes and inconsistent referral patterns but make the most of (and possibly increase) the strengths of each approach. Examining the strengths and overcoming the weaknesses of both biomedicine and Chinese medicine, as well as considering where they can be combined, results in practical beneficial effects for actual patients. As improvement of patient care is the foremost reason for integration, case reports are a good way to demonstrate how these types of interactions can offer a window into how health care decisions are carried out for actual patients. And as with any case reports, significant findings can be used to support the design of research studies to evaluate effectiveness on a larger scale.

The Strengths and Weaknesses of Biomedicine

Due to a number of clear strengths, biomedicine has become the predominant paradigm for medicine today. Biomedicine is well-known for its ability to exert an immediate and powerful influence on health. Technology has led to the capability of refining and concentrating medications to affect biochemical processes. Surgical procedures continue to be refined to remove, replace, or repair structural damage. Because of the aggressive and directed approach to biomedical treatment, this type of therapy might generally be considered the best approach for patients with certain acute conditions or those in emergency situations such as cardiovascular events, stroke, and traumatic accidents.

Biomedicine is also based on substantial knowledge of anatomical physiology. The current level of detail in our understanding of physiology is astounding. For example, scientists have created models that explain specific biochemical transformations at the molecular level. From a communication standpoint, because individuals in the West are taught to understand nature through a mechanistic vocabulary, medical practitioners often explain illness and treatment in biomedical terms, and patients generally use the vocabulary of science as their entry point to communicate their symptoms and understanding their health state. This dominant system of communication about health problems is a useful frame of reference.

Technology is another strength of biomedicine. Diagnostic tools developed in the 20th century such as x-ray, CT scan and MRI allow physicians to see more clearly into the body to identify structural causes of illness and to also verify if the structural problems appear to be resolved after treatment. Through blood and urine tests, physicians can monitor levels of dozens of biochemicals, where too much or too little may indicate a lack of homeostasis and either an explanation or warning of present or future symptoms. The biomedical technologies for treatment are also highly evolved. To introduce a biochemical compound into the body beyond the straightforward topical and oral routes, subcutaneous and intramuscular injection or intravenous drip can be used.

Biomedicine also has its share of weaknesses. Iatrogenesis is one major problem that concerns the adverse effects of the powerful and immediate types of therapy such as surgery and medication. The strength and quickness of biomedical treatments often have costs in terms of adverse effects, which can at times be as debilitating as the original condition to be treated.[9] Medications have long lists of potential adverse effects, and although surgeries can be used to resolve severe problems, they certainly involve risk, and recovery may take time. Medical error is the third leading cause of death in the US. In 2004, a university hospital found that 36% of over 800 consecutive patients had an iatrogenic illness, with 9% of all patients admitted experiencing disabling or life-threatening incidents.[10] The Institute of Medicine estimated in 1999 that between 50,000 to 100,000 deaths annually were due to errors in hospital care in the U.S.—and since then this number has been recognized by numerous sources as being too low.[11]

Beyond adverse events, another weakness of the biomedical approach is the difficulty in creating effective biomedical treatments for chronic and complex diseases. David Dingwall, Canadian minister of health in the mid-1990s, noted in a speech that "the promise of more and better drugs has not often translated into better health. As disease patterns have shifted from the acute to the chronic, from infectious to degenerative, progress in dealing with chronic illness has been more limited."[12] Broadly speaking, biomedicine's reliance on anatomical physiology may also be limiting, especially for conditions where the physiological details are not completely understood. And for conditions with multiple causes, finding a one size fits all biomedical monotherapy may not be a viable option. Finally, the knowledge base of biomedicine may shift over time with new research. There have been numerous cases where a medication which was proven or assumed to be safe in the past may prove years later to have harmful side effects (for example, hormone replacement therapy and the cox-2 inhibitor rofecoxib).[13]

There are also aspects of biomedicine that are both a strength and a weakness. Consider for example biomedicine's heavy reliance on biomarkers as a measurement of disease. In many cases this is a strength of biomedicine, but patients who have symptoms but no abnormal lab findings to explain them have frustratingly few options for treatment. In addition, there are a number of conditions where biomedicine has not yet discovered a clear biochemical or physiological cause for a specific symptom or condition. For these syndromes that have unclear etiology and pathogenesis, it is difficult to find a starting point for designing successful treatment.

The Strengths and Weaknesses of Acupuncture

Acupuncture has its own set of strengths. The observation skills of traditional acupuncture, passed down through many generations, are a considerable strength of the practice. Acupuncturists use observation to detect the nature of a disease through palpating the acupuncture channels and the pulse, by observing the color, sound, odor, and emotion, inspecting the tongue or the external ear, or through other traditional modes of observation. Any irregularities when palpating the meridians can enable practitioners to explain and predict certain phenomena beyond what we understand about biomedical mechanism. These types of observations are particularly useful in cases where patients either have symptoms not yet explainable through biomedicine, or where patients' lab values are normal despite their symptoms. In these instances, observations may help to provide a direction for treatment, whereas biomedicine might be at a loss for treatment planning due to the lack of diagnostic information. A patient may also benefit from traditional diagnosis and corresponding treatment if his lab values are borderline but still within normal range. Rather than waiting until the lab values put him over the threshold, he might decide to pursue acupuncture treatment as a preventive measure.

Another strength of traditional acupuncture is related to its paradigm and process — the association and grouping of symptoms to diagnose a pattern of disharmony.[14] Even without biomedical testing an acupuncturist can obtain enough information through the patient interview and physical exam to formulate an effective treatment by considering the symptom as occurring within a context. A related strength of acupuncture is its emphasis on treating the root of the disease, addressing the pattern observed rather than simply a single symptom. The recognition in Chinese medical thought concerning the interconnectedness between different parts of the whole — not only interactions between organs but also connections between points on acupuncture channels — means treating the root not only addresses the individual symptom but may also resolve other concurrent symptoms.

Yet another strength of acupuncture lies in the variety of its treatment principles and techniques. Some acupuncture patients require supplementation to boost weak areas, or harmonization to restore balance to multiple functions. The directed actions of biomedicine rarely involve using an agent to directly nourish the body, as much of biomedicine is allopathic in nature where drugs have the opposite effects from the symptoms (bacterial infections are treated with anti-bacterials and hypertension is treated with anti-hypertensives). Further unique treatment options exist within acupuncture; for instance, in the source/*luo* point combination where strength from one channel is borrowed to make up for weakness in another, or in orthopedic style needling, where an acupuncture needle is an excellent tool for releasing tight muscles and freeing local channel obstructions.

Chinese medicine also has its share of weaknesses. One of these is the multifactorial approach to treatment. A patient might receive body acupuncture and auricular acupuncture simultaneously, followed by a course of Chinese herbal medicine, and recommended *qigong* therapeutic exercises as self-care. When the patient improves, no one knows exactly which therapy is effective, and for this reason, Chinese medicine is not as efficient as it could be.

A second weakness of acupuncture is the subjectivity of results. Because acupuncture was practiced before the technological development of outcome measurements such as blood pressure and urine and blood tests, the observational skills routinely used can often detect qualities that are not easily measured, and inter-rater reliability may be compromised. Comparing baseline to post-treatment outcomes may be difficult solely using traditional diagnostics. Indeed, quantitative patient-centered biomedical outcome measures may be used integratively to supplement qualitative outcomes.

Table 1. Strengths and Weaknesses of Biomedicine and Acupuncture

	Biomedicine	Acupuncture
Strengths	Powerful immediate actions and effects Basis on anatomical physiology Diagnostic tools and measures Treatment technologies	Pattern differentiation and treating root and branch Model of the channel system Recognition of interconnectedness between the parts and the whole Treatment principles and unique therapies
Weaknesses	Adverse effects and iatrogenesis Poor treatment options for many chronic conditions Heavy reliance on biomarkers Shifting knowledge of physiology and therapeutic interventions	Multifactorial approach to treatment Subjectivity of results

Applying Strengths and Reducing Weaknesses

This list of strengths and weaknesses of both biomedicine and Chinese medicine is by no means complete, but considering how to use these lists will give us a place to begin contemplating how to truly integrate acupuncture and biomedicine. There are three main ways to use these lists of strengths and weaknesses. Strengths can recommend one approach as a treatment priority for referral. Combined strengths can synergistically be linked to increase quality of care. Strength in one approach can be used to shore up a weakness in the other. With each combination, the goal should be to increase at least one of six characteristics of good health care as determined by the Institute of Medicine: efficient, safe, effective, timely, patient-centered, and equitable.[11]

The first strategy is referral — to choose the modality that prioritizes the use of the most relevant strength based on the immediate needs of the patient. For example, a hypothetical patient may be struggling with lumbar pain, with recent sudden onset symptoms of urinary and fecal incontinence. Biomedical knowledge of neuroanatomy may help a care team expedite spinal surgery to prevent these losses from becoming permanent. Another patient with lumbar pain who is not judged to be as urgent may wish to pursue a course of gentle, noninvasive acupuncture for relief before resorting to surgery. If successful, this modality could potentially save the patient time, effort, and financial resources. Each case takes into account one or two key strengths from the above chart to improve the quality of care.

A second type of integrative strategy involves synergy — the concurrent or successive use of biomedical therapy and acupuncture treatment to take advantage of the strengths of both. Instead of merely using acupuncture as an alternative chosen in the event that biomedical care is not adequate, or vice versa, we may find some instances where a combination of biomedical and acupuncture treatment together would result in a boosted effect than from using either alone. In one case report, a 13-year-old girl with postherpetic neuralgia received an immediate course of intravenous acyclovir, with a regimen including stellate ganglia block, medications, and transcutaneous electrical nerve stimulation (TENS) for relief. A subsequent 10-week course of acupuncture followed that resolved her residual persistent facial pain and nausea. In this instance, the immediate effects of an anatomically-based biomedical treatment was combined with acupuncture's channel system model to improve the patient's overall health.[15] In another case report, a 65-year-old patient with traumatic brain injury from a motor vehicle accident received acupuncture treatment at point GV 26 (*Renzhong*) and bleeding therapy at the 12 *jing*-well points after emergency craniotomy and brain surgery, resulting in regained consciousness and the ability to begin rehabilitation after three weeks of treatment.[16] The sequential interventions were essential in restoring the patient's consciousness and function.

The third type of integration strategy involves leveraging — using a strength of one branch of medicine to make up for a weakness of the other. A hypothetical instance where a strength of acupuncture is used to counteract a weakness of biomedicine in practice can be seen in patients who receive acupuncture treatment to address the side effects of chemotherapy regimens. For a cancer patient, if the white blood cell count remains acceptable, chemotherapy may continue as scheduled without interruption. In this situation, the strength of acupuncture in this case is its ability to supplement and strengthen the patient's vitality while addressing the biomedical weakness brought on by the adverse effects of chemotherapy treatment. Using acupuncture to treat any pharmaceutical adverse effects or postsurgical pain can address weaknesses in the biomedical paradigm.

Conversely, using objective biomedical measures and diagnostic tools to verify the benefits of acupuncture treatment can be helpful to make up for the weakness of subjective evaluation in acupuncture outcomes. One example of combining a strength of biomedicine with a strength of acupuncture concerns a patient treated for male infertility; after a three-month course of acupuncture, the patient's sperm parameters were increased by between 34% and 50%.[17] In this case, biomedicine's strength was the diagnostic technology to verify that the treatment was of use. Without the technology of semen analysis, ascertaining the success of the acupuncture in achieving its goal would only be reflected in the occurrence of a healthy birth of a child 40 weeks later. In light of the few effective biomedical treatments for this condition and due to the presence of many other variables, biomedical testing helps to isolate this specific variable to verify the effectiveness of the treatment. While in this instance biomedical involvement is not an intervention, it aids in helping to ascertain progress with a long-term treatment goal. It is important to remember, however, that biomarkers are merely correlated with the disease state; improving the biomarker doesn't necessarily mean that the suffering has been remedied, and if the lab value does not change that does not necessarily mean the acupuncture is not having a positive effect.

Many examples illustrate this idea of practical integrative medicine. One patient with carpal tunnel syndrome may receive a course of laser acupuncture to solve her problem, using the strength of biomedical technology together with the strength of the acupuncture meridian system as an alternative to surgery. A second patient with sensitivity to anesthesia medications may receive acupuncture before a biomedical procedure for pain prevention. A third patient might notice improvement in heart arrhythmia after acupuncture treatment verified through biomedical EKG testing. In these and countless other cases, improved health is achieved using a truly integrative approach. Investigating the combinations more thoroughly and establishing a practical database of integrative case reports may help allopathic practitioners to make more consistent and appropriate referrals, while also making acupuncturists aware of situations where collaboration can benefit patient care.

Conclusion

If the aim is to integrate biomedicine and acupuncture into a combined approach that genuinely uses the strengths of both sides, care must be taken when using the word 'integrative.' As discussed above, the first conception of 'integrative medicine' involved absorbing CAM treatments into the dominant biomedical system of care, possibly without regard for the complementary paradigm. The case report approach, where strengths and weaknesses are juxtaposed to improve patient care, can be used instead to consider a variety of therapy combinations — everything from acupuncture combined with chiropractic care to biomedical care combined with aromatherapy. In every combination, it is important that the essence and paradigm of the approach is not compromised. This concept — known as resonance — should be reflected in the discussion of the case report, and the therapy described should reflect the underlying system or paradigm of that therapy.[18] Discussion of the thought processes involved in care decisions is essential to preserving accumulated knowledge with the goal of better medicine. If case reports are the earliest type of evidence that reveals new ways of thinking, case reports as practical demonstrations of care should influence the directions for research and education in integrative medicine.

Preparing for a truly integrated health care system in which strengths of both modalities are preserved as much as possible, requires that initiatives be taken in research, education and for its use in actual practice. Any treatment combinations which seem to be effective in a case report may be tested clinically to verify that the combination's effectiveness is not limited to just one case. In education, students can be made aware of the pitfalls surrounding integrative medicine in order to ensure the authentic practice of acupuncture. And in practice, acupuncturists can be attentive to identifying situations where the strengths of acupuncture may be juxtaposed with other therapies to increase favorable outcomes. Practitioners can apply case report recommendations in their practices to see if these strategies tend to result in similar improvements. In this way, case reports can help inform integrative practitioners in making patient care decisions. In the end, improvement in actual patient care should lead the way the integrative structure is built.

As more case reports accumulate, patterns and commonalities in groups of cases will emerge, pointing to a deeper understanding of how to combine these therapies in practical ways. If common threads are able to be identified in these successful combinations, integrative medicine education can also have a practical side beyond simply relying on attitudes, values, and referrals. In this way, resourceful practitioners who will be able to develop new collaborative ideas may fuel the future of integrative medicine.

Part II

As discussed in the first half of this book, case reports are essential for the preservation and development of traditional acupuncture. Historically, case report records have provided a window into the shifts and divergences in the practice of acupuncture and have demonstrated how different interpretations and viewpoints can lead to an effective treatment application. Because of the difficulties in subjecting traditional practice to modern validation through clinical trial and mechanism studies, the role case reports can play in evidence-based medicine is greater than might first appear. Examining how successful integrative care has been accomplished in single cases can serve as examples for stimulating both research hypotheses and practice models based on clinical results.

The reasons for writing case reports are therefore clear. The second part of this book serves as an instructional manual, describing how to draft a case report clearly and with rigor. Chapter 6 outlines some of the reasons to choose a particular patient as a case report subject, whether it concern an unusual diagnosis, an innovative treatment, or complex symptom picture. Chapter 7 reviews in detail the types of information necessary for a comprehensive patient interview in clinic to provide the necessary raw material for writing a case report. After considering what should be included for thorough charting, the next several chapters outline how to write the contents of each individual section of the case report. These are not presented in an order reflecting the sequence of the completed work, but rather in an order that makes the writing process easier and more straightforward. Chapter 8 describes how to organize that chart information into a compelling narrative that describes the case, including patient presentation, diagnosis, treatment, and outcomes. Chapter 9 lays out the requirements for writing a case report introduction that includes background material about the condition, while Chapter 10 explains how to structure the discussion and conclusion, where the significant points of the case are explored and summarized. Chapter 11 includes instructions on how to create a title and abstract, with guidance on references to be made throughout the paper. Finally, Chapter 12 describes the process of assembling and preparing a case report for publication and outlines a rubric that can be used to evaluate the completeness and quality of a case report.

CHAPTER 6

Reasons for Writing

WHAT MAKES A patient a good subject for a case report? The first patient who comes to mind might be one who had a particularly good outcome — perhaps a faster response than expected or a more complete recovery than what might be considered typical. However, readers will be interested in reading cases beyond only those reporting good results. For example, case reports describing adverse events can be useful in warning other practitioners away from making similar mistakes.

When choosing a patient for a case report, consider the audience. Some readers may be researchers, looking for protocols. Some may be practitioners of other modalities, looking to make an acupuncture referral. But many will be acupuncturists looking to learn new skills and techniques and deepen their understanding of traditional theory. Knowing how to choose the case report subject may determine whether the paper will be useful for readers. A textbook case describing a standard treatment for an average patient that corresponds with established practice guidelines is not likely to be published.[1] But a case report about a patient with average results but describing a relatively unknown technique not yet published in any textbook may be more interesting and useful in practice. The same could be said about relating the exploration of a course of treatment for a patient that reflects both setbacks as well as successes in order to model ways for the reader to manage difficult cases. Complex patients with multiple patterns may also be good subjects for case reports, so a reader can learn to recognize what aspects of diagnosis are most important or what treatment direction is best to start with. Even a case where a patient did not respond well may reveal useful knowledge, depending on the analysis of the case. Journal editors look for original and though-provoking case reports that have sufficient educational value to serve as a novel learning activity for the reader.[2]

Choosing a Patient for Your Case Report

As you consider which case to write about, rather than asking 'what case can I write about,' ask 'what case would a practitioner want to read about?' and 'how would this case report help the reader become a better practitioner?' Too many case reports conclude with a statement stating something like 'acupuncture may be an effective treatment for condition x. More research should be done.' This type of throwaway statement is an unfortunate consequence of many case reports having largely been written for the purpose of generating hypotheses for potential clinical trials. Writing cases only with this type of conclusion in mind overlooks practice-related benefits of reading case reports. The case report can be a window into the mind of a practitioner: describing how the treatment was chosen may give the reader insight into the medicine beyond simply providing a set of points that might work for a similar patient. The following categories may help you to consider what types of cases might make for interesting reading.[v]

Patients with an unusual condition or presentation

For a rare condition where clinical trials might not be possible, a case report may be the only evidence we have of successful treatment anywhere. For example, Stiff Person Syndrome (SPS) is a very unusual neurological diagnosis, and there is no consensus on how to treat this disorder biomedically. Since so few people have ever been diagnosed with this condition, it would be nearly impossible to conduct a clinical trial. In 2005, a case report was published which detailed the successful treatment

---

v Appendix B lists examples of case reports in each category.

of a patient with SPS; at the time it was one of the few literature sources available on the treatment of that condition.[3] New conditions for which clinical trials have not yet been designed can be fertile ground for case reports. Other cases that may fall into this category include unusual presentations of symptoms where the patient's physician has not been able to find a suitable biomedical diagnosis.

Patients with recalcitrant conditions

Some patients seek acupuncture as a treatment modality if they do not respond to standard biomedical care or if their conditions are difficult to treat biomedically. If a patient receives benefit from an alternative therapy for a condition where biomedicine tends to have little success, a case report documenting the interaction may provide unique insights not available from other information sources.[1] These biomedically challenging diseases are often complex and multifactorial. For example, case reports have been written on post-traumatic stress disorder, schizophrenia, complex regional pain syndrome, and fibromyalgia — all challenging conditions because either mechanism of these diseases are not adequately understood, because the presentation of each individual may be different, or because no biomedical treatment has yet been devised to address them consistently. Describing a treatment course that successfully addresses a recalcitrant condition may not only help readers learn how to treat difficult patients but may also result in referrals to acupuncture from other medical professionals.

Unique treatment approach

Many acupuncturists have learned special techniques from teachers and mentors that have not yet been published. Some practitioners, through their own insights or by combining ideas from multiple sources, have even come up with effective new strategies of their own. If you have a case that follows this description, explaining this unique approach in the context of a case report is a good way to disseminate this knowledge quickly. Even if theoretical material has already been published on the approach you used, writing a case report can make the technique easier for a reader to apply. Examples of this type of case include discussion of new point locations, applications of less common styles of acupuncture, and the use of techniques for different problems than originally intended. Innovations in point stimulation may be included here, such as the application of essential oils topically to acupuncture points. The condition in this type of case may be commonplace; it is the novelty of the treatment that makes the case report stand out.

Challenging diagnosis

A case report that models problem-solving is useful for practitioners to read. It may be difficult to make sense of a patient with scattered signs and symptoms; in these cases, it may be difficult to identify a pattern to focus treatment on. Is there a key observation (e.g., regarding the tongue, pulse, or abdominal finding) that may provide a clearer direction to correct diagnosis? In some cases, the diagnostic pattern is unexpected or not listed among common patterns for the condition at hand. Another scenario that can help readers through a challenging diagnosis might be a case where an initial incorrect diagnosis resulted in either no improvement or even worse, an exacerbation of symptoms, but after the diagnosis was changed, the patient responded favorably. Describing the thought process in retrospect and pointing out the underlying pattern of thinking can help a reader avoid making those same diagnostic mistakes in the future. Explaining a challenging differential diagnosis in a case can help readers better understand the process of Chinese medicine.

New medical technology

The field of acupuncture is full of creative practitioners who apply new materials and devices to traditional practice. This technology may include new acupuncture instruments, such as laser and electroacupuncture machines, as well as unusual herbal applications, such as acupoint injection. If the literature explaining a new technology is lacking, case reports can be well-situated to fill that gap. As substantial time and energy are required to publish a book on new techniques or technologies, a case report may be an easier and quicker way to explain how and when to consider applying them. Even if sufficient literature describing the use of a new technology has been published, case reports can help to document success in practice and offer guidance in its correct application. Furthermore, case report writers may offer any additional recommendations which can further enhance effectiveness.

Application of traditional theory

Many statements in classical acupuncture theory are written in a manner that can be difficult to interpret. Multiple interpretations may be possible for a number of reasons, including changes in the Chinese language from classical to modern usage. Translation into languages other than Chinese brings new challenges and new interpretations as well. Multiple interpretations and perspectives are possible, depending on the school of thought. The value of a particular interpretation depends on whether that interpretation informs effective treatment, which is demonstrable through a case report. For example, the instructions on treatments addressing the secondary vessels (sinew, *luo*, and divergent channels) are vague in most textbooks, and specific examples on how and under what circumstances to perform these treatments would be very informative. Similarly, the usage of Sun Simiao's Ghost point category is poorly understood; a case report describing a patient's presentation, the techniques employed, the number of points used, and the outcomes would add significantly to the literature. Clarifying a possible meaning of any challenging concept through demonstration of a successful case can enhance the potential of acupuncture.

Reporting adverse events

As with any modality, we must be aware of possible adverse events. Writing and reading cases on this subject can prevent potential problems with similar cases encountered in practice. Although acupuncture is generally quite safe, and serious adverse events are not common,[4] case report literature is heavily biased towards the publication of adverse acupuncture events. Despite acupuncture having a superior safety record, the percentage of adverse event articles in acupuncture greatly exceeds the percentage of adverse event articles in biomedicine.[vi] This may be due to a number of potential reasons, including publication bias and the fact that licensed acupuncturists are rarely published in biomedical journals. Therefore if a case involves an *unusual* adverse event, it is important to bring awareness to other practitioners through a case report, but if it involves a common adverse event that acupuncturists are well aware of, consider choosing a different topic.

Unexpected effects (benefits) of treatment

Patients often come to acupuncture with a constellation of symptoms and complaints. When acupuncturists address an underlying health imbalance and the chief complaint resolves, other symptoms may also resolve. For those who are familiar with acupuncture, this type of case report helps to reinforce the importance of good diagnostic skills and can remind us of relationships between seemingly unrelated symptoms. For example, in one case report, a practitioner observed a marked reduction in restless leg syndrome after *guasha* scraping technique was used to treat shoulder pain, hypothesizing that *guasha* has circulatory effects that extend beyond musculoskeletal pain conditions.[5] Another case report describes an infertility patient seeking treatment for headaches. After her acupuncturist diagnosed and treated a Phlegm pattern and disharmony in the Kidney channel, not only did her headaches resolve, but she also was able to conceive, as theoretically the underlying imbalances were also causing her fertility problems.[6] For a biomedical reader, this type of case may introduce the concept of pattern recognition and holism in the practice of acupuncture.

Confirmation of something already known

Although a case report that confirms textbook knowledge may not be as illuminating for an advanced reader to learn from, if you are a student struggling

---

vi For example, a PubMed search for "case report" returned over 2 million articles, with only one out of the first 50 results reporting an adverse event (2%). By comparison, a similar search for articles under "acupuncture case report" revealed 2111 case reports with 25 of the first 50 results reports of adverse events (50%).

to find a case, there is nothing wrong with choosing one that demonstrates the traditional process of acupuncture diagnosis and treatment, even if it does seem to follow all the expected turns. Although the discussion points in this type of case may be less salient (making the case report unlikely to be published), what matters is how the process of writing helps develop clinical thinking. Many beginning case report writers realize while writing that their treatments included several unnecessary points, leading them to strive towards leaner and more elegant strategies for future patients.

Integrative medicine

As opportunities to collaborate with practitioners outside Chinese medicine increase, the potential for improvements in patient care increases as well. Acupuncturists involved in an integrative practice with practitioners of other modalities may write about examples of successful collaboration that can be used to build a model for integrative care. As discussed previously in chapter 5, using multiple modalities in an integrative way can improve patient-centered care by educating both acupuncturist and non-acupuncturist readers on knowing not only when to refer, but also how to combine strengths of one modality to overcome weaknesses in another. For example, when using the treatment principle of supplementation in acupuncture to support the vitality of a patient as he endures adverse effects of a chemotherapy regimen, one of the strengths of acupuncture (unique treatment principles) can be used to bolster one of the weaknesses of biomedicine (adverse effects). Piecing together examples of how to combine modalities effectively in practice can create a better basis to support the specific roles of acupuncture logistically in an integrative medicine setting. If relevant cases are not documented, a valuable opportunity to establish the place of acupuncture in this transition is lost.

Multiple treatment approach comparisons (or combinations)

The variety of acupuncture treatment approaches being taught today is exciting, but with the number of options available, the choice of which style to apply in any given situation may not be clear. Perhaps a patient did not feel much improvement with the first approach that was tried but then responded very well after a change in needling style. Writing about successes of the various approaches helps the field sort out the strengths of each and can lead to more efficient treatment. Understanding the variety of practice styles also has implications for educational design, i.e., how to train acupuncturists, what material to focus on within each approach, and how to specialize according to style, biomedical domain, or condition.

Case series

Finally, the category of case series deserves some attention. A case series is an article that reports on two or more cases. A case series might also be a way to demonstrate that more than one case had positive outcomes with a common approach, making the likelihood of the cause and effect relationship slightly more compelling. The cases may be similar in that patients with the same biomedical diagnosis or same traditional pattern diagnosis may be compared. Alternatively, an author might choose to include more than one case to contrast the cases for an informative point, related to diagnosis, treatment, or management (e.g., contrasting two different styles of acupuncture in addressing the same chief complaint). A writer can even tease out the key differences between similar patients that could be used to demonstrate how a principle might be applied differently depending on the presentation. Writing a case series paper tends to be more involved than a single case report for a number of reasons; several cases must fit into a limited word count, each case typically is presented in an abbreviated form while making sure to include important details, and additional comparison between the individual cases must be done, often by creating charts and tables.

This list of reasons for choosing a case to discuss is by no means exhaustive, but they all share the common goal of conveying ideas that can help a reader become a more astute practitioner, as well as provide benefit to patients.

Case Reports and Promoting Good Medicine

Beyond considering the reasons described above for choosing a case, it may be useful to think about how writing and publishing your case may benefit medicine as a whole. In 2001, the National Academy of Sciences specified six qualities that are essential to good health care: *safe, effective, patient-centered, timely, efficient, and equitable.*[7] While these qualities are certainly used as touchstones to improve the delivery systems of medicine, they can also apply to the individual therapies administered to any patient. When deciding which cases might be most apt for reporting, consider whether any of these essential qualities of medicine may be improved if the outcomes of your case were typical.

Safety is a growing concern in medicine due to the ever-increasing development of new biomedical treatments. A case report may have the potential to increase safety through one of two means: the first through warning acupuncture colleagues through case reports of possible adverse effects of acupuncture, and the second through demonstrating how acupuncture in an integrative capacity may potentially improve the safety index of biomedical treatment. With regard to the latter, acupuncture may be used in lieu of biomedical therapies associated with greater risk or in combination with biomedical therapies known for their adverse effects. If in combination, the acupuncture treatments in your case may occur before, after, or concomitantly with the biomedical treatment.

Effectiveness can be defined as the quality whereby therapy is successful in producing intended outcomes in a real-life clinical setting. The ability to produce a desired result in ideal circumstances — known as efficacy — is testable by clinical trial using large groups for statistical analysis. This differs from effectiveness, which is the actual goal of medicine — the ability to successfully solve the health problems of an individual patient. The acupuncture case report is a complex observation that attempts to understand effectiveness in a single patient. In some cases, effectiveness can mean eliminating the symptom completely. However, complete resolution of a condition is not a fair expectation of effectiveness for all conditions, since treatments in biomedicine such as regular insulin injections or long-term hypertension medication may be considered effective as maintenance despite ongoing care. Reading cases on how practitioners can apply skills and knowledge in a way that arrives at an effective outcome can be useful for both beginning and continuing education. Case reporting is an ideal research method that can help us understand the application of a complex system of thought, requiring a flexible use of that knowledge which can be adapted to individual patients.

Patient-centered care focuses on meeting a patient's individual preferences, needs, and values.[8] By nature, the case report has the potential to describe how a therapy or intervention can be specifically tailored to an individual patient's wishes and expectations. Case reports proposing standardized treatment protocols or those omitting descriptions of how treatment fits a particular patient ignore the ideal of patient-centeredness. Traditional acupuncture theory maintains that each patient is unique, and patient-centered care extends beyond mindful dialogue to influence the design of an acupuncture treatment course. Case reports that describe the way the acupuncturist accounts for individual characteristics in diagnosis, together with explanations of how treatment is personalized, enable readers to understand the logic behind a patient-centered intervention.

Timeliness refers to the quality of medicine that delivers the right therapy at the right time. This quality of medicine is largely dependent on the systems of delivery. Timeliness of care in acupuncture could be improved in at least two ways through case reports: by increasing the speed of appropriate referrals and by describing effective approaches so that practitioners will have a better idea about how to most quickly affect positive change. Case reports written for non-acupuncturist readers may educate them on when to refer; referrals may also happen between acupuncturists when one has not had specific training in the skill set described in a similar case that shows good results. The second way to increase timeliness is through individual case reports that compare the effectiveness of two styles or two methods on a single patient. In the future, analysis of aggregated case reports may be able to improve timeliness through guiding clinical choices. With increased search capabilities via electronic databases, it will become more possible to do this type of analysis so long as adequate case reports exist.

Efficiency in medicine concerns the use of the best medical resources to produce the best outcomes. Part of efficiency is cost-effectiveness, but it also can involve the best use of materials (e.g., hospital beds). Efficiency is often a function of systems of care; for example the adoption of technology (e.g., electronic charting methods) might appear to be one way to save time and energy while improving communication, thus ostensibly improving patient care.[9] Case reports can suggest that acupuncture may increase efficiency in patient care by comparing cost-effectiveness between the acupuncture intervention and the biomedical. For example, if a patient can avoid an expensive procedure such as lumbar surgery after a course of acupuncture, a writer may choose to include a statement describing the amount of money and resources that might be saved. Case reports aimed at describing pre-surgical patients who may benefit from acupuncture referrals may potentially lift the financial burden from patients as well as possibly eliminate the physical and emotional costs of recovering from surgery. If a case report highlights an acupuncture technique that delivers quicker results than the typical acupuncture care, this may also save resources.

The final essential characteristic which good medicine strives for is *equity*. This ideal specifically addresses the inequality in care with regards to race and ethnicity but could also be expanded to inequality of care and outcomes caused by disparities in age, gender, sexual orientation, income, or geographic location.[10] Researchers are working to detect where and how these inequalities manifest, to understand why these inequalities exist, and to create ways to reduce them.[11] Although very few previously published case reports focus on inequalities in the medical system, writing one on this topic may draw attention to those inequalities and any possible suggestions for rectifying them. These may take the form of identifying poorly treated conditions endemic to a specific region (e.g., using moxibustion to treat tuberculosis in Africa) or suggesting ways to provide acupuncture to individuals without medical insurance or financial resources (e.g., explaining treatment styles used within community acupuncture clinics).[12] A case report may inspire socially-minded practitioners to replicate these efforts to improve equity in medicine.

CHAPTER 7

The Patient Chart

BEFORE WRITING BEGINS, it is important to make sure that the information required to write a complete and detailed account of the patient has been properly collected in the patient chart. Many beginning writers find that their chart notes are not quite thorough and that certain key points were not covered adequately in the patient interview. For example, perhaps range of motion was not recorded before the first treatment began, or perhaps the patient was not asked at the initial intake how many headaches she experienced per week. When writing a case report, any omissions in the chart notes become very clear. Writing a narrative based on an incomplete record becomes a more difficult endeavor and readers will not be able to easily benefit from your work. It is therefore crucial to understand how to perform a comprehensive and accurate patient intake so all necessary raw materials are available when writing a rigorous case report. You may end up collecting more material than you need and not all of the patient intake information will be necessarily included in your case report. However, it is important not to miss any important details you wish you had asked about. A thorough intake is required to support the choices made in diagnosis and treatment.

Patient History

The primary goal of the intake is to gather enough information about the patient to make a clear diagnosis. Begin by asking about the chief complaint, which may be identified as either a key symptom or group of symptoms that the patient is experiencing. The details of the chief complaint are sometimes recorded in the medical chart as the History of Present Illness (HPI) or History of Presenting Complaint (HPC). In biomedical practice, OPQRST is a mnemonic commonly used by biomedical practitioners to gather information about a patient's chief complaint, where each letter represents a specific aspect of the patient's chief complaint (onset, palliation/provocation, quality, region/radiation, severity, and time).[1] While OPQRST is a useful tool to ensure that you have covered the range of questions, each individual element within the mnemonic might not always be applicable to every situation and condition.

Onset

When, and under what circumstances, did the symptoms begin? The answer to this generally includes an aspect of time (e.g., onset occurred three months prior to the start of treatment) and in some cases may include a clear aspect of causation (e.g., onset occurred after falling off a ladder). The onset may be gradual or sudden. If the cause was clearly a specific event, it is necessary to mention this due to its possible influence on the direction of your treatment. If the symptom is a flareup of a chronic condition, it is important to include a remark about when and how the symptom first began, the nature of the flareups (approximately how often they occur, how long they tend to last, what approaches have been tried in the past), and when the most recent flareup began.

In some cases, a patient might not necessarily associate the onset of the symptoms with a particular event, but with some creative questioning and practitioner experience, a related event might be drawn out. Go beyond asking 'When did you start to have these symptoms?,' where the response is often a minimal answer defined only in terms of time. Instead, ask open-ended questions such as 'Can you describe how did these symptoms began?,' or 'Was there anything significant that changed in your life around the same time?' These may yield more fruitful answers. For example, a patient might not remember that six months before the onset of his diarrhea he changed his diet and began consuming a daily soy protein shake for breakfast. Another patient might not have associated the onset of her asthma to the week after the passing of an elderly relative.

An extension of the onset element is the progression of the condition over time. Some patients experience exacerbation and remission of their complaints over time and this is important to capture as well. One example is the patient with lower back pain, whose emergency laminectomy five years before reversed extremity numbness, only to have gradual numbness return within the several months before seeking acupuncture. Another example is the patient with a recent recurrence of Bell's palsy, where she had recovered completely from her first episode in her late 20s after a difficult childbirth. If the patient had previously experienced the same or similar symptoms in the past, this too should be noted in the chart.

Palliation/Provocation

Did any factors seem to provide symptom relief? Were there factors that seemed to cause symptoms to worsen? One possible factor is *weather* conditions — for example, hot and dry weather made your patient's joint pain feel better, or cold and damp weather made his joint pain feel worse. Some specific *activities* may exacerbate a symptom — a patient might have felt his knee pain worsen only while running, or perhaps he experienced neck pain only after two hours of working on a laptop computer. On the other hand, other specific activities may provide relief — soaking in a hot tub reduced a patient's lower back pain. Specific *positions* may relieve or intensify pain — a patient's hip pain tended to improve while seated or worsen when standing. Or perhaps the symptoms began with specific *movements* — shoulder pain occurred when using a hair dryer. *Foods* can also trigger certain symptoms such as gas and bloating after eating raw foods. What kinds of *medications* made the symptoms recede and at what dosage? How much did it help and for how long? What other types of *treatments* has the patient tried, and what has offered relief? During your intake, some of these factors may help you to differentiate the condition, determine the cause, and come up with lifestyle or dietary suggestions that will help to alleviate the symptoms.

Quality

Noting the quality of pain conditions may add support to your pattern differentiation.[vii] Most commonly the patient is asked to describe a sensation, whether dull and aching, sharp or numb, burning and/or radiating. If the patient reported numbness, be sure to specify whether she was suffering from true numbness (lack of sensation) or paresthesia (a pins and needles type of sensation). Sometimes a type of discomfort may be related to an internal condition (e.g., chest pressure for a cardiac patient). The experience of a patient's mental-emotional symptoms may also be described in terms of quality; for some patients panic attacks might feel like a strong rushing upwards sensation whereas others might feel numbness or tingling. Some patients with depressed mood might feel general fatigue and sleeplessness; others might tend towards a loss of appetite and digestive discomfort. These qualities or types of sensations can heavily influence acupuncture diagnosis and treatment and should be duly noted.

Region/Radiation

This element is also often associated with identifying the region of discomfort and any area where the symptom radiates to. For example, pain in the right elbow may originate from the lateral epicondyle but spread slowly upwards to the acromion, or it might radiate from the medial epicondyle to the palmar aspect of the fourth and fifth fingers. The location of the pain would then help to support your choice of acupuncture points later in the treatment section. Region can also be applicable in symptoms that are not related to pain such as numbness, paresthesia, lack of strength, and dermatological symptoms (ranging from subjective itching to more objective lesions). Locations and regions of these types of symptoms may be further defined during the physical exam.

---

vii If the patient's complaint did not involve pain or discomfort (e.g., infertility) the quality aspect of the condition is moot.

Severity

Most commonly considered in the context of pain management, severity can be gauged in the initial intake by asking patients to rate their pain. This can be done using a Numerical Rating Scale (NRS), where patients are asked to rate their pain with a number from 0–10, with 0 being no pain and 10 being the worst pain imaginable. Another useful pain scale is the Visual Analog Scale (VAS), where a horizontal line is drawn on a piece of paper, with the left end of the line representing no pain and the right end of the line representing extreme pain. The patient is asked to indicate a spot on the line where their current pain lies. The position of the X is then measured and compared before and after a treatment course. Yet another useful pain scale is the Wong-Baker FACES® Pain Rating Scale that presents a series of drawings used in conjunction with numbers for a patient to indicate his or her pain level.[2] This scale is often used in pediatrics but can be used for all ages.

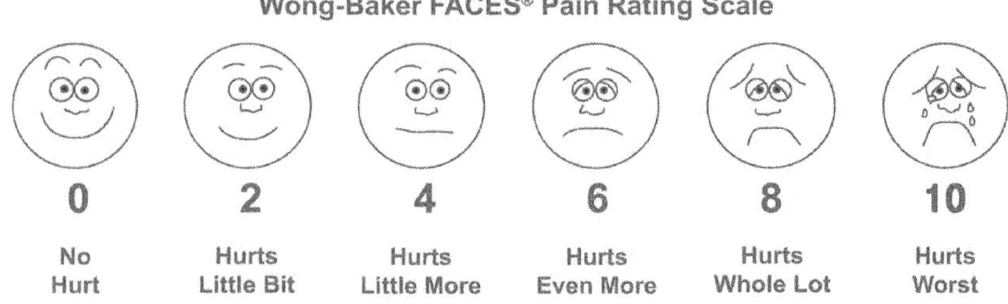

©1983 Wong-Baker FACES Foundation. www.WongBakerFACES.org
Used with permission. Originally published in Whaley & Wong's *Nursing Care of Infants and Children.* ©Elsevier Inc.

Verbal Rating Scales are also an option for evaluating the severity of symptoms. A severity Likert scale uses words instead of numbers to describe the intensity of symptoms (e.g., 'choose one of the following: none, mild, moderate, severe, very severe'). When a patient's symptoms are not related to pain, it is possible to adapt the scale to gauge the patient's level of distress or the severity of the symptom being considered (for example, 'on a scale of 0-10, how would you rate your level of anxiety?').

Sometimes a patient may find it difficult to choose a number on the scale because the symptom intensity varies so greatly over time. For these cases, the patient may indicate a range rather than a specific number. It may be also useful to include a comment on how that range manifests over time. For example, a patient felt headaches over the previous week, where most of the week the pain was at a constant level of 2/10, with only one morning when the pain level spiked to 7/10.

Timing

A variety of factors make up the time element. The first type of question has to do with *duration*: How long does the symptom usually last? Answers can range from continuous to the shortest moment. If the symptom is intermittent, the second type of question related to timing concerns *frequency*: How often does the symptom occur? Possible answers may include that the symptom is constant, occurs several times per minute, or is experienced on average once per month. A full reporting of frequency and duration of irritable bowel syndrome, for example, may include that the patient experienced flareups approximately once every four to six weeks, with each typically lasting for one week before subsiding. Sometimes frequency does not occur on any sort of regular basis (e.g., a patient's sciatica flareup occurred once last year, but three times during the previous year). Likewise, duration may also be variable for recurrent flareups. Beyond duration and frequency, *recurrence* is important to note: is there any predictability to the timing of the symptom? Perhaps there was a particular time of day when the symptoms are likely to appear (e.g., neck pain in the morning on waking). Sometimes this may also be a time of the week (e.g., anxiety on Mondays). Time of the month may also be possible (e.g., migraines at the time of the menses) as well as seasonal recurrence (e.g., allergic rhinitis in the springtime).

In an intake, OPQRST items do not have to be asked in any particular order. Usually, after the practitioner opens with 'How can I help you?,' the patient may identify the symptom, but then might jump straight to addressing aspects of Quality and Region/Radiation. A seasoned practitioner will listen to the patient and ask questions that follow naturally from the patient's lead. It is essential that all OPQRST items have been sufficiently covered before wrapping up the chief complaint portion of the interview.

Ten Questions/Review of Systems

Questioning about the chief complaint is followed by a review of the patient's overall health. This often comes in the form of the TCM Ten Questions or a more biomedically-based Review of Systems (ROS). The holistic information in this section can help to support a traditional diagnosis. The TCM Ten Questions are not actual questions, but are topics that can provide information about the patient's overall condition which can be used to inform diagnosis and treatment design. The topics for the ten questions vary slightly depending on the reference text, but the most common areas of questioning are chills and fever, sweating, head and body, thorax and abdomen, food and taste, stools and urine, sleep, deafness and tinnitus, thirst and drink, and pain.[3] In an acupuncture student clinic setting, a clinical supervisor may require all ten areas to be covered. In published articles, only relevant items that are out of normal range are generally reported.

The Review of Systems approach is a way to spell out the patient's general health condition and tends to be used more often in biomedical journals. In this format, the patient's bodily function is reviewed system by system. These include respiratory, cardiovascular, digestive, genitourinary, endocrine, dermatological, musculoskeletal, neurological, eye, ear/nose/throat, hematologic, and psychiatric. Some interviewers ask about social and family history as part of the interview. If you choose to pursue the Review of Systems style of inquiry, be sure to familiarize yourself with the types of questions for each system. For example, 'How is your cardiovascular system?' may be challenging for a patient to answer fully, whereas if you were to ask specifically about palpitations, chest pain, and shortness of breath, these types of questions may result in more revealing answers. If you do choose to use the Review of Systems, be aware of questions that are relevant to traditional diagnosis which may have been overlooked. You may find that you need to supplement your Review of Systems questions with traditionally oriented questions. Does your patient tend towards feeling hot or cold, is there unusual sweating or perhaps subcostal tension? These types of questions may be instrumental in forming a traditional diagnosis.

Beyond obtaining information, the patient interview process enables you to establish rapport. Avoid inadvertently asking the same questions twice, once for the chief complaint, and then again for the Ten Questions/Review of Systems, as this will signal to the patient that you are not listening. Good interviewing skills include obtaining all of the information that will be useful for making an accurate diagnosis and creating an effective treatment.

Physical Exam

The physical exam includes making note of relevant objective findings. Being able to compare the state of the patient before and after a treatment course will add greatly to the rigor and professionalism of your case report. Some objective findings may be in the biomedical domain, while others may be more traditional in nature.

Measurements can be used at the start of treatment to objectively gauge the severity of the condition. For cases relating to pain and musculoskeletal dysfunction, objective findings may include range of motion, sensitivity to palpation, and analysis of movement. Range of motion may be measured with a goniometer; using this simple and inexpensive tool can add significant credibility to any conclusions made in cases involving movement impairment. Documenting a reduced range of motion at the start of treatment provides a baseline for comparison at the end of the treatment course, and periodic measurements made during the treatment course may also help you to gauge progress. A flexible tape measure can be used to measure girth of atrophied or edematous limbs.

Posture and gait may also be observed as an element of severity. Speed and smoothness of movement may be observed objectively using functional measures such as the Timed Up and Go (TUG) test and the Six

Minute Walk Test (6MWT). The TUG test is a measurement of the time in seconds it takes for a patient to stand from sitting in an armchair, walk three meters, turn around, walk back to the chair, and sit down. The person wears usual footwear and may use a cane or walker if they do normally. This test measures mobility for patients with pain, stroke disability, and vertigo.[4] The 6MWT measures the distance a patient can walk during a six-minute interval on a hard, flat surface. The patient is self-paced and may rest as needed. This test may be useful to evaluate mobility for patients with knee or hip arthritis as well as fibromyalgia, heart and lung conditions, and stroke.[4]

Objective measures can also be used to record the region/radiation of the condition, particularly for neurological and dermatological symptoms. A region of numbness can be delineated by using a sharp/dull test, and paresthesia symptoms can be described in location and area. For example, a patient's foot and lower leg numbness (lack of sensation) presented in a sock distribution pattern to a line 15 cm proximal to the ankle joint. When examining dermatological complaints, photographs may be very useful in documenting size, shape, and color of a lesion. For example, a patient with eczema presented with one bright red dry round lesion at 5cm in diameter at the left popliteal fossa. Photos taken before and after a treatment course may also provide a visual comparison. Before photographing a lesion, place a 10cm ruler next to it to provide a helpful frame of reference.

Orthopedic tests may also help to establish a useful baseline for conditions related to musculoskeletal pain. While this aspect is not related specifically to any element of the OPQRST, comparing a patient's function using manual muscle testing or other orthopedic tests can be useful to inform point selection and treatment technique for orthopedic style acupuncture, offer a reference point for comparing the patient's post-treatment function, and signal red flags if referrals are necessary. Relevant tests can be found in orthopedic exam textbooks.[5]

Other objective biomedical findings may include results from lab tests and radiological exams, whether these were performed by you or by the patient's biomedical physician. Lab markers relevant to the diagnosis can be included in the write-up but only if you also have access to post-treatment lab values for comparison in your results section. The more significant the change in the lab results, the stronger your case will be, especially if you intend to submit your paper to a biomedical journal.

The physical exam in acupuncture includes traditional methods of looking, listening, smelling, and palpation.[3] When looking, one might perceive details of appearance, facial hue, observation of the *shen* (spirit), or description of the tongue. One might notice spider veins on the Kidney channel at the ankle, a swelling at SP 9 (*Yinlingquan*), or red irritated eyes, each of which can be used to help support a particular underlying pattern diagnosis. In Five Element acupuncture, observing the odor (putrid, rancid, rotten, pungent, burnt) and sound (groan, shout, laugh, weep, sing) can be useful information. In TCM, listening to the volume of the voice or the forcefulness of the cough may also be useful.

In a number of different styles of acupuncture, palpation is an important aspect of diagnosis. The pulse presentation should be described using specific terminology that appears in an accepted text. For example, bowstring and wiry[6] and robust pounding[7] are acceptable pulse descriptors. Using terms that are not a part of accepted pulse systems is not recommended; for example, using 'puffy' to describe the patient's pulse is meaningless in a case report because the reader will not know what puffy feels like, and no standard interpretation exists for that term. Beyond the palpation of the pulse, palpation of channels and points may also be a useful diagnostic tool. A variety of approaches to abdominal palpation (especially in Japanese styles of acupuncture) can be useful in informing diagnosis. Palpation of channels, either for unusual warmth, coolness, moisture, hardness/softness, or pressure pain, can also be good information and in some styles may be key in diagnosis and/or treatment.[8] The front *mu* (alarm) points and the back *shu* (associated) points are also traditionally palpated for diagnostic purposes. Nodules and indurations can be described by hardness, size, and location.

In the treatment room, the patient history and physical exam are discrete sections of the information gathering process. In reality, in some cases it may be useful to continue asking questions during and after the physical exam if the findings warrant further investigation. For example, if a patient's chief complaint is hip pain, and a physical exam revealed an ankle surgical scar that was not discussed during the patient history, it would be important to ask about the ankle surgery to ascertain its relationship to the hip pain. Certain types of diagnosis may lend themselves to concurrent questioning and palpation, such as the palpatory diagnosis of channels in the style of Wang Ju-Yi[8] and in Kiiko Matsumoto style acupuncture.[9]

Outcomes Measures

Outcomes measures are used to gauge progress during the treatment course and potential effectiveness at the end of the treatment. The term itself is somewhat of a misnomer because while these measures are certainly used at the end of a treatment course to evaluate outcomes, they must also be taken before treatment begins in order to have a point of comparison. Only by establishing a clear baseline are we able to evaluate the success of the treatment and determine what the benefits of treatment may have been. Some outcomes measures noted below may overlap with details in the OPQRST sequence as well as in the physical exam, but this sequence may miss some useful points of comparison. You may therefore find it useful to consult this factor list as you finish your intake. Doing so will ensure that you have recorded all of the baseline items and have a clear comparison when writing your case report. These factors may also be helpful in preparing written reports for health insurance companies that require frequent progress updates.

In general, the patient chart should track the overall trajectory of symptom relief, not just the baseline and final outcome. Sometimes the patient can feel worse before feeling better, sometimes there can be a delay in the onset of relief, and sometimes the patient can feel better for a short time before the symptoms return. These may change over the trajectory of treatment: if the patient's symptoms felt better for one day after the first treatment, and two days after the second treatment, then three days after the third treatment, this response can be useful to document so your reader can learn what kind of results might be possible following the treatment course. It is worth noting that asking questions related to these symptom factors may be useful for encouraging hesitant or skeptical patients in clinic. Some patients may not realize they have improved, especially when the changes are incremental. If you can clearly point out specific ways a patient has improved, he or she may be more likely to continue treatment.

In recent years, Patient-Reported Outcome Measures (PROMs) have been increasingly used as instruments to measure patient-reported outcomes, particularly in conditions that are difficult to quantify. PROMs usually take the form of questionnaires that are filled out by patients before and after treatment. They are designed to evaluate the change in the effect of symptoms on daily activities, quality of life, and mood (e.g., anxiety/depression).[10] PROMs are often used as quantifiable outcome measures in clinical trials, where it is impossible to report qualitative data on each individual in large sample sizes, for use in statistical analysis. Case reports generally do not require statistical analysis. For this reason, questionnaires are not as useful in individual case reports. Although PROMs may be useful in reducing bias, and including them in your case report may certainly help bolster the validity of your outcomes, relying on them as a sole measure of improvement may cause you to miss opportunities to capture the individual patient's state. When details are reduced to numbers, important descriptors can be lost that may be useful for the reader in judging if and how to modify the described treatment to match a similar patient in their own clinic.

In a patient-centered evaluation of a symptom or condition, seven symptom factors may be helpful to consider for outcome measures. These factors are *intensity, frequency, duration, location, stress capacity, resilience,* and *medication use.*[viii] Keeping these in mind during the intake may help you more clearly define your initial baseline and allow you to create better comparisons when it comes time to writing the case report. Familiarizing yourself with these not will not only infuse your case report with

---

viii Some of these factors are described in Bauer[11] and I have added other useful factors.

adequate detail but also document your results for chart review, help establish rapport with your patients, and sharpen your clinical analysis.

Intensity is similar to the idea of severity. For pain, one can attempt to quantify this subjective quality with either a numeric rating scale (NRS) or a visual analog scale (VAS). Alternatively, one might choose to define the severity by using a verbal description of how much the pain interferes with activities of daily living. For example, a patient may indicate at the start of treatment that he experienced shortness of breath and pain walking up the stairs before treatment, but he was able to do so without trouble after treatment. For complaints that are not pain-related, it is also possible to describe how the patient's life activities are hampered. A patient's anxiety kept her from leaving her home or from going out with friends on weekends; after seven acupuncture treatments, she was able to go on outings on several occasions. At the end of your treatment course, you can compare the results to the baseline to show clear functional improvement.

Frequency of symptoms is the second factor that can help you to understand the degree of improvement. Comparing how often symptom flareups tend to occur before and after the treatment course can be helpful. An insomnia patient might wake at 1 a.m. seven nights per week as a baseline, but then at the end of the treatment course, he would wake only one night per week. Note that frequency improvements don't always have to be in the decreasing direction; if a patient with constipation has only two bowel movements per week, increasing this to daily bowel movements by the end of the treatment course is an improvement. At times, the frequency of symptoms may be more complex. For example, a patient might experience episodic cluster headaches where she might have a severe headache every day for two weeks, with an average of three episodes per year. Improvement might involve fewer headaches within the episode or fewer yearly episodes.

Duration of symptoms is another outcome measure that can change with treatment. Note if the symptom is intermittent, when the symptom occurs, and how long it lasts before settling down. For example, a patient might report at the intake that each monthly migraine headache would last for three days, but after a series of treatments, the headache typically would resolve within one day. A patient suffering from irritable bowel syndrome might typically experience spells of urgent diarrhea lasting one week, but after a series of treatments, flareups would only last two days. Even if the frequency is the same, improvement can be verified if the duration is reduced.

The *location* of patient's complaint can shift throughout a treatment course. Sometimes an area can recede, increase, or move, and improvements can be noted in your chart when they occur. For example, a patient with sciatica might have initially experienced radiating pain down the leg to the ankle but after three treatments indicated that the pain only radiated to the thigh. Or a patient's eczema lesion which was 10 cm across started to lighten and recede, and after six treatments, only a small patch 2 cm in diameter remained. Changes in the measured region of complaint may indicate improvement.

Stress capacity is another way to measure outcomes. Many complaints have stress-related factors that make them more noticeable. For example, if a patient at first reported that he felt hip pain after standing for 10 minutes, but after a series of treatments was able to stand for 30 minutes before the onset of pain, this would be considered improvement. Sometimes stress capacity can be defined by time (e.g., how long can the patient sit or stand) and sometimes it can be defined by distance (e.g., how far can the patient walk or run before symptoms start). Recording baseline stress capacity will enable you to understand after treatment is completed how much improvement was achieved regarding a patient's function in the presence of triggering factors.

Resilience, also called rebound capacity, is another component that can be used to define the baseline. The resilience factor is defined as the amount of time it takes for a person to recover from an exacerbation. Imagine you have a patient with occasional hip pain who runs marathons. After each marathon, her hip pain flares up, taking about four days to settle down and feel normal again. After treatment, she may find that even though the post-marathon pain may be similar in intensity, the pain flareup would dissipate by the following morning, indicating an increase in resilience.

Finally, *medication use* is an important outcome measure. Perhaps your patient initially took six tablets of ibuprofen per day, and after a few treatments, needed only two tablets per week. Even if she reported that her pain level did not change, the reduction in medication is notable. (Note that unless prescribing is within your scope of practice, you should not recommend to the patient any modification of the medication dosage, including discontinuing a medication. During the course of treatment if you believe that the medication might no longer necessary, reach out to the patient's prescribing physician to discuss this or recommend that the patient ask the physician directly.)

You may choose to track any of these categories more closely at various points in the treatment course, rather than simply before treatment begins and after the final treatment. Evaluating these factors frequently, even at every treatment if appropriate, can help you monitor clinical progress and decide when to change or modify your treatments.

Performing a thorough patient interview, physical exam, and evaluation not only enables you to collect the information you need for your case report, but this is a necessary skill for thorough diagnosis and patient-centered treatment design. Practitioners who write case reports for the first time as professionals often find that their clinical skills and logic are sharpened by the process of putting their thoughts into written form. Knowing what information needs to be gathered is the first step in writing a thorough case report. The next chapter discusses how to weave this information into a smooth narrative, the case description.

CHAPTER 8

Case Description

THE CASE DESCRIPTION section should be a clear and chronological account of what happened in your case. Because all of the information required for writing this section is contained in your patient's chart, this section is one of the easiest places to start writing. Your task is to create a smooth narrative that describes what happened clinically. The full case description includes three main portions. First, describe the patient's history. What was the patient's chief complaint, and what were the findings from the physical exam? Second, describe the diagnosis and treatment. What diagnosis was given, what treatments were applied, and what was the rationale? Third, describe the outcome: What happened, and how do the patient's results compare to the initial baseline condition?

As the case history describes the patient's state at the time of the initial intake, this whole section should be written in past tense. The material should also be presented logically to allow the reader to follow the practitioner's thought processes as the case unfolded. Keep in mind that the case report should be written as an academic paper. Use complete sentences and paragraphs, making sure that your writing does not take on the appearance or feel of chart notes. Note that clinical signs are either present or absent, not positive or negative. When using an acronym, place it in parentheses after the first time it appears in the text (which is most often in the introduction section). Make sure any acronym you use is commonly accepted by checking its use in the established research literature. Avoid using chart shorthand notation and abbreviations (e.g., use 'blood pressure instead' of 'BP'), and acupuncture jargon (e.g., use 'vacuity' or 'deficiency' instead of '*xu*'). Translate all theoretical pinyin terms except for *yin, yang,* and *qi*. A number of acupuncture journals by convention capitalize terms that refer to traditional concepts. For example, Liver would be capitalized when describing traditional theory (e.g., Liver *qi* or the Liver channel), whereas liver should start with lowercase when referring to biomedical concepts (e.g., liver enzymes or liver transplant).

To guard the patient's anonymity, do not include certain information such as the patient's name, initials, identity numbers, biometric data, or geographic location (name of town or city where the patient lives or received treatment). Specifying an area smaller than a state could enable a reader to discover the identity of the patient and therefore should be avoided so as not to compromise confidentiality.[1] The mention of a specific geographic region may be included only if it is relevant to the patient's condition or care. For example, if a patient who surfed regularly on the Oregon coast developed an aching hip pain exacerbated by exposure to cold and damp weather, it would be acceptable to use the 'Pacific Northwest.' If you include an image (including x-rays, laparoscopic images, ultrasound images, or pathology slides) you must obtain written consent from the patient as well as remove any identifying text.[ix] In general, avoid quoting the patient, as paraphrasing provides for much smoother reading (one exception to quoting a patient directly is if you are reporting on the patient's mental-emotional state).

Do your best to write simply and succinctly. Readers for whom English is not their first language may have difficulty understanding language that is abstract, poetic, or literary in style. When writing a case report, it is helpful to read another case report written in the same style for reference. This model case report can potentially provide a pattern for you to follow. If your case report is an academic assignment, ask your instructor to provide a model case in the desired format. If you intend to submit it to a journal, check the journal's *Instructions for Authors*, and read several previously published cases to get a sense of the knowledge base of the

---

ix See Appendix C for more details regarding patient consent.

readership as well as understand that journal's conventions for how case details are presented.

Patient History

Start with the health history by describing the patient (age and gender) and their condition: 'This case report describes a 79-year-old male with a history of chronic cough,' or 'The patient was a female 50 years of age, with symptoms of hot flashes, night sweating, and insomnia.' To keep your writing focused, if the patient has multiple complaints, you may decide to choose the most prominent one as the chief complaint. The opening statement of this section may alternatively state the biomedical diagnosis given by a physician: 'A 60-year-old male with a diagnosis with peripheral neuropathy in both hands sought acupuncture treatment for his discomfort.' While it would be appropriate to reveal the biomedical diagnosis here, it is best to avoid prematurely discussing the traditional acupuncture diagnosis until you have laid the groundwork for its support.

The chief complaint does not necessarily need to be presented in OPQRST order, and the information from the patient interview may be mixed with information from the physical exam as long as it reads smoothly. If any of the OPQRST items or initial evaluation items are not relevant, they may be omitted (e.g., describing the Quality component for a patient whose chief complaint is infertility). Similarly, although subjective and objective information is usually recorded in distinct sections in the chart, mixing material from these distinct sections within the narrative of a case report is acceptable if it helps make the section flow smoothly. For example, in the interest of keeping chronological order, a description of the patient's experience of the condition may be interspersed with findings from biomedical evaluation in the interest of keeping chronological order: 'The patient was a 35-year-old female with a chief complaint of dysmenorrhea. The onset was gradual, starting three years before as a deep aching pain starting two days prior to the onset of menses. After a particularly painful period during the first year, she was diagnosed by her physician with a 2 cm uterine fibroid. The pain became more severe as time went on, and the timing of the pain began to occur up to five days prior to menses. By the second year, ultrasound showed that her fibroid size had increased to 3 cm.'

Beyond mixing the description of the patient's subjective experience with findings from biomedical testing, data from the physical exam may also be incorporated into the narrative to aid in comprehension: 'The patient, a 35-year-old male, had sudden onset right shoulder pain which he associated with lifting weights at the gym during the previous week. The pain was dull and aching and awoke him from sleep at least once per night. His range of motion was limited to 100 degrees of abduction, 75 degrees of external rotation, but with full forward extension. Tender points existed at the anterior deltoid, which felt ropy in texture.'

One exception to presenting information in chronological order involves elements of the history that may inform how the condition began or developed. If a past event may have influenced the trajectory of the disease, it may be recounted after describing the presentation at the time of the initial patient interview. For example, after first describing relevant clinical details about a 50-year-old patient with irritable bowel syndrome, it would be reasonable to then state that the patient visited India at age 25 and was treated successfully for intestinal parasites at that time.

It is acceptable to juxtapose the facts of the case to indicate that a patient associates his condition with a potential contributing factor: 'A 25-year-old male patient presented with frequent temporal headaches that had developed gradually during the previous six months. Shortly before the onset, he started a new job working nights, and to stay alert, he consumed a caffeinated energy drink before each shift.' Alternatively, if you as the practitioner are making this type of connection independently, you may want to hold off on making any assertions about the cause and reserve the etiology (cause of symptoms or conditions) until after you present your diagnosis. Finally, be sure to include any aspects of the patient's social or family history relevant to the case.

After thoroughly describing the chief complaint, the next section of the case report should review the patient's overall health. This review should be written in paragraph form rather than as a chart or table. The two most common approaches to the review are the TCM Ten Questions format or the biomedical Review of Systems. (If you are unsure about which one to use, model your writing after previously published case reports in the relevant journal or ask your instructor.) Some journals may omit the Ten Questions or Review of Systems in the interest of reducing word count.

For the Ten Questions or Review of Systems, normal findings may generally be omitted, but all relevant information that could influence your diagnosis and treatment must be presented. For example, if your patient tended to have one solid bowel movement per day without difficulty, this is within normal limits and does not need to be reported in your case report. One exception is to include a normal finding if it will later help you to rule out a competing diagnosis. For example, if a patient experienced fatigue, frequent urination, and felt cold but did not have loose stools, it may be useful to include the fact that the stools were well-formed. This supports Kidney *yang* deficiency as a diagnostic pattern and allows you to rule out Spleen *yang* deficiency. In some published journals, the Ten Questions or Review of Systems section is significantly abbreviated, including only items that affect the diagnosis and treatment. If the patient's overall health is normal and there are no significant findings, one sentence will suffice: 'All other body functions were within normal limits,' or 'The patient denied any other significant health concerns.' Lastly, if the patient has unique concerns, preferences, or expectations that affect the treatment, these should also be included.

Biomedical evaluation may add to the rigor of your patient history. Include lab values that may be relevant, but only if you have the pre- and post-treatment measurements to compare. Blood pressure and temperature do not need to be included unless they are relevant to the condition. Tables and charts may be used in your written case report if you have multiple relevant lab values to track.

If not already included in earlier material, relevant findings from the physical exam may be added here. The wide variety of acupuncture approaches make use of a similarly wide variety of observation skills and findings. Pulse readings, tongue appearance, and palpatory findings are among the information presented. A finding should be included only if it is relevant to the patient's diagnosis and the style of treatment. For example, if a patient with ankle pain presented with tension at abdominal point CV 12 (*Zhongwan*), but that point was irrelevant to your diagnosis and treatment, you may omit. In a case report, you do not need to report everything that is written in the chart; good technical writing contains all the necessary information without being superfluous.

As you write your case description, you may find you omitted asking some relevant questions during the initial patient interview. One common experience of new writers is that the process of writing a case report alerts them to what parts of their patient interviews are inadequate. Ultimately, this process stimulates them to greatly improve their patient intakes. Be as specific as you can in the interview, and you will have plenty of information for your case description. Lack of information leaves holes in your logic, which may not be as noticeable while treating your patient, but these gaps become much more evident as you describe your thought processes when writing.

Once you have written the case description, reread to make sure that your language is concise and your account complete. Only then will you be able to present and support a diagnostic assessment.

Diagnosis

After all relevant information has been laid out in the case history, your next task is to make sense of that information and develop a diagnosis. As discussed earlier, acupuncture has evolved into a multifaceted study with a great variety of perspectives. One challenge in writing a diagnosis section is to make the thought process clear to the reader even if he or she may not have been exposed to the style of acupuncture used in your case. If the style of acupuncture is not well known, it may be helpful to include a paragraph about this lesser-known approach in your introduction section (discussed in chapter 9) rather than attempt to explain the style and the diagnosis simultaneously. Do not introduce any new information concerning the patient history; while it may be tempting to do this in order to make the writing more compelling, it will make your case report appear disorganized.

Diagnosis in TCM style acupuncture

For a TCM diagnosis, both a disease diagnosis and a pattern differentiation should be included. The disease diagnosis is the symptom or group of symptoms and is not a biomedical diagnosis. It consists of a two or three syllable Chinese word, such as *tou tong* (headache), *wei zheng* (atrophy syndrome), or *shan qi* (shan disorder). These are commonly covered in chapters of internal medicine textbooks or as topics covered in a TCM pathology class.

A TCM pattern is an imbalance that is supported by a number of related signs and symptoms. Each pattern you diagnose for your patient must be supported by at least two, but preferably three (or more) relevant signs and symptoms. For example, a patient might be diagnosed with Heart Blood deficiency if she presents with insomnia, palpitations and a pale tongue. A common TCM error is to base a pattern on one symptom alone. For example, if a patient only has palpitations, the palpitations may be due to any of a number of patterns, including Heart Blood deficiency, Heart *yin* deficiency, or Heart fire. It is impossible to choose one without additional signs and symptoms to corroborate. If the diagnosis is challenging or if there are competing patterns that were difficult to rule out, explain the decision-making process that allowed you to arrive at your final diagnosis.

Another common error in TCM diagnosis is the use of a biomedical pathology or condition to support a pattern. For example, novice practitioners have often used hypertension as one of the support symptoms for Liver *yang* rising. A patient with hypertension may have a number of primary patterns, and many hypertensive patients do not have Liver *yang* rising patterns (conversely, many patients with Liver *yang* rising do not have hypertension). To keep the integrity of Chinese medicine intact, a modern biomedical condition cannot be used to back up a TCM pattern.

A third common error in TCM diagnosis is the attempt to use an element of etiology or pathogenesis to support a pattern. For example, a writer might state that a patient has a diagnosis of a Kidney *yang* deficiency pattern because (among other symptoms) the patient is elderly. While old age may contribute to the development of Kidney *yang* deficiency, it may not be used to support the diagnosis because it is not a sign or a symptom. Two other common examples of this error are using 'poor diet' as a symptom to support a Spleen pattern or 'stress' to support a Liver pattern.

The TCM pattern diagnosis for your patient may be fairly complex. A single patient might have signs and symptoms that span several patterns. For example, your patient may have Heart Blood deficiency, Spleen *qi* deficiency with dampness, and Kidney *yang* deficiency concurrently. Before deciding to include all of these in your report, you may want to make sure that all of these patterns are related to the treatment of the chief complaint. If the above patient has a chief complaint of chronic diarrhea, you may choose to leave out the Heart Blood deficiency if unrelated or relegate it to a secondary pattern.

Another way to check your patterns retrospectively is to examine your point choices to see whether each pattern was addressed in the treatment. For example, if no points were used to address Kidney *yang* deficiency in the above example, and the patient's condition does not relate to the Kidney *yang* deficiency pattern, you may omit that pattern in the diagnosis. In such a case, you may bring up the Kidney *yang* deficiency pattern in the discussion if you wish to mention that the Kidney pattern, although present, did not play a strong role in this particular case.

Diagnosis in non-TCM style acupuncture

The methods of diagnosis vary as much as the styles of acupuncture do. Take care to follow the principles of whichever style you have used in your treatment to make sure that your diagnosis is clear and remains authentic to the original style. For example, in Five-Element Acupuncture, possible diagnoses include Aggressive energy, or Wood Causative Factor; and in Kiiko Matsumoto Acupuncture, they might include *Oketsu* or Adrenal imbalance. With any non-TCM diagnosis, you must provide a justification that is consistent with that style of acupuncture. For example, for a patient to be diagnosed with a Wood Causative Factor, the patient would need to have a green color, shouting sound, rancid odor, and an imbalance in the anger emotion. If a patient is diagnosed with a Kiiko Matsumoto style *Oketsu* pattern, a palpatory finding such as tension in the left ST 25–27 area should support that diagnosis.[2] Citations supporting the logic behind the diagnosis should be provided, especially for diagnoses that are not standard TCM.

Lastly, the diagnosis in your case report must match the style of treatment you have used. The diagnosis must be drawn from the case description and lead logically into the treatment principles. The styles of diagnosis and treatment should match. It would be confusing to offer a diagnosis of TCM Heart Blood deficiency followed by a description of a Five-Element Aggressive Energy treatment. The reader might come away with an incorrect assumption that everyone with Heart Blood deficiency should be treated for Aggressive Energy (or vice versa).

Treatment

The goal of the treatment section is to not only to accurately describe how treatment was designed and administered but also to provide justification for your treatment choices. This reasoning, also called rationale for treatment, must be stated clearly. Readers should be able to understand the logic you used to design your treatment so they can apply your ideas to similar patients they may encounter.

Complexity of treatment may vary greatly. The simplest case might describe the use of one acupuncture point for one treatment only, whereas the more complex cases might use a large number of acupuncture points, along with shifting treatment goals and multiple modalities through various stages of therapy. Standardized criteria for the descriptions of acupuncture protocols within clinical trials have been created,[3] but the descriptions of acupuncture interventions in case reports may differ slightly. Because protocols in clinical trials are written for strict reproducibility, every detail must be clearly defined. However, because real-world acupuncture interventions may involve dozens of treatments with variation from treatment to treatment, reporting every point used at every treatment may be cumbersome.

One challenge to reporting on treatments given is to know what to include and what not to include. If the patient in your case was seen and treated in the clinic just once, the details of the treatment could be reported exactly as performed. Describing the treatment in all its detail may be straightforward if there are few treatments and few points, especially if the same treatment points are consistently applied at each treatment. However, in cases when the acupuncture treatment course involves a large number of sessions, not every detail of your chart notes may be worth including. Also, if the point prescriptions used in numerous treatments vary, reporting every detail may make the case unreadable. In these cases, it would be acceptable to summarize, while ensuring that the rationale behind treatment choices is clear, any specific techniques are described in detail. Most cases fall somewhere between these two extremes where treatment choices are adjusted for changing conditions. A practitioner reading the case report would be able to reasonably reproduce or modify the treatment to match their own patient. The goal is to write as precisely as you can without sacrificing the readability of the case.

The treatment section should start with the treatment principles or goals. In TCM, each pattern should correspond to a treatment goal. For example, for a patient with Heart Blood deficiency and Spleen *qi* deficiency with Dampness, the treatment goals should include nourish Heart Blood, supplement Spleen *qi*, and resolve Dampness. If you divide the treatment focus, concentrating on your Chinese herbal treatment to nourish Blood and your acupuncture treatment on supplementing *qi*, that would be useful to note in the treatment principles. Non-TCM styles have their own treatment goals and you may want to consult textbooks on those styles to accurately describe the goals of your treatment. For example, for Japanese meridian therapy you could state that root treatment was performed to address pattern imbalances present at the time of treatment, noting all patterns that presented during the course of treatment and specifying which one(s) were most common.

The treatment principles should be followed by a summary of the overall treatment course that describes the number of sessions performed within a specific timeframe. For most cases, that period spans from the first time you met with the patient and performed the initial intake to the final treatment when the patient was released from care. For example, you might record the patient received ten treatments over the course of four months, once per week for eight weeks and then two additional treatments at once per month to maintain the improvements achieved.' In some cases, you may wish to focus on a shorter window of treatment. For example, if the patient's condition resolved after eight treatments, but he continued treatment to relieve a second unrelated condition, you may wish to focus on the first eight treatments in your case report, with your final results reported after the eighth treatment. If you are writing your case about an acute condition which occurs in the midst of a longer treatment course for another issue, you may focus on the shorter window. The treatments you report on should also be in succession (e.g., you should not just report on treatments 1, 5, and 8).

Specify the needles used for treatment, including length and gauge, manufacturer, and country of origin in parentheses. Brand and country of origin should be noted for moxa. Equipment should be identified by listing the manufacturer, model number if applicable, and the country of origin in parentheses.

The short simple case[x]

After describing the treatment principles, the treatment course, and the equipment, the next step is to list the points and their functions. For a simple case, where the treatments were either few in number or consistently similar from session to session, a single chart is most convenient to present these treatment points (for example, see Table 2).

Table 2. Acupuncture Points

Acupuncture Point	Action/Indication[4]	Technique
ST 36 *zusanli*	Benefits Stomach and Spleen, supplements *qi* and Blood	Moxa
SP 6 *sanyinjiao*	Nourishes *yin* and Blood	Even
PC 6 *neiguan*	Calms the Spirit	Even
LI 10 *quchi*	Harmonizes the Stomach and frees Intestine, supplements *qi* and Blood	Supplement
CV 12 *zhongwan*	Harmonizes the Stomach and fortifies Spleen	Even
LV 3 *taichong*	Courses Liver *qi*	Drain

List each point along with its pinyin name (in italics) in the left-most column, the action and indication that you were considering when you chose the point in the second column, and the technique in the third column. For action/indication, include only actions and indications that are relevant to your usage in this case. For example, in Table 1, while the point ST 36 (*Zusanli*) has many potential actions and indications, the relevant ones that are listed are brief and to the point. Be sure to cite sources that support the actions and indications of the points.

Using a table to compile this list makes the information easy to absorb by your readers. If points were added to the primary point prescription but not used at every treatment, you must indicate that either in the chart or in the subsequent paragraph. If your needling depths were unusual, you will need to mention that either in an additional column in the chart or in the text following the chart. Some channel names may have multiple abbreviations (e.g., Triple Burner could be abbreviated TE, TW, or TB). Check the conventions of the journal you are considering in order to choose its preferred abbreviation.

In the body of the text, it would be useful to describe what type of supplementing or draining technique was used (e.g., lifting and thrusting or rotation). Some writers choose to add a fourth column with the depth of needling; another option is a simple statement indicating that needles were inserted at standard depths according to a specific text (with supporting citation). If you use adjunctive techniques, you will need to specify details, including the nature of the moxa technique (e.g., thread moxa, warming needle), the materials and/or methods of cupping technique (e.g., fire, moving, plastic, glass) as well as any other necessary details so the reader can approximate your treatment procedures for a similar patient they might see in clinic.

If your treatment course is lengthy but still fairly consistent from session to session, it may be useful to group the points into two separate charts, with the first including key points used at all treatments and the second including auxiliary points used as needed depending on the presentation on the day of treatment. If you use more than one chart, be sure to describe each one appropriately. An average of the number of points used or a range may be appropriate to include in the text. As above, the chart should also include point actions and indications, technique, and depth of needling for each point. Additionally, you may wish to include a table indicating the number of treatments where each point was applied, as seen in Figure 1:

Figure 1: Frequency of Acupuncture Point usage

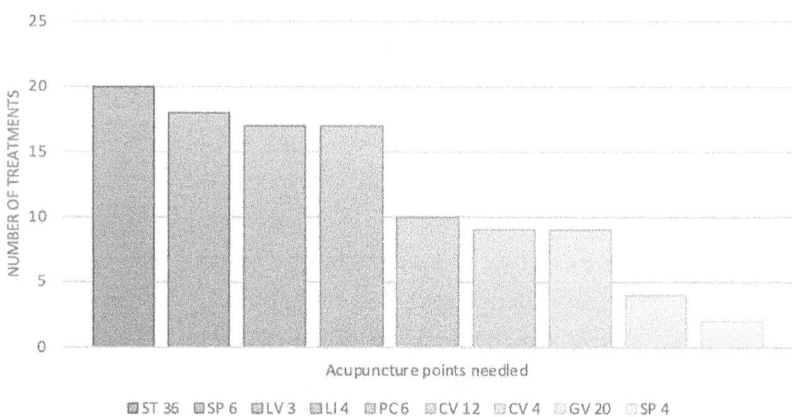

Other modalities used[x]

Other modalities administered to the patient related to acupuncture should also be described in this section. Some of these may be based on the same traditional theory as acupuncture, such as Chinese herbal medicine, traditional Chinese dietary therapy, *qigong* exercises, and acupressure. Since those techniques relate to traditional theory, describing the rationale will help readers better understand whether these recommendations would be useful for their own patients. If you make any recommendations that are not acupuncture-related, these must also be reported, including (but not limited to) specific muscle stretches and orthopedic exercises, meditation techniques and breathing exercises, exercise routines, western dietary recommendations or supplementation, sleep habits, and any other recommendations for self-care between visits.

---

x Because Chinese Herbal Medicine is so common it is discussed in Appendix D.

Describing your reasons for making these recommendations may be useful (with references), as will brief explanations of any that might require instruction. (If lifestyle recommendations were made by other practitioners and did not originate from you, these should be included in the case history rather than the treatment section.)

Outcomes

After the treatment details have been provided, let the reader know how your patient responded. If the treatment course was short, a description of the patient's condition after the treatment course will likely be sufficient. If the treatment course was lengthy, it may help to describe the patient's response over time. How you describe the improvements and changes will depend upon the case: some patients have a marked response to the first treatment, whereas other patients respond gradually over many treatments. You may report as many or as few outcomes as are relevant to your particular case. If your case does not show specific improvement in a given parameter, you may wish to omit that in the outcomes section. Tables or graphs may help the reader visualize the results more easily.

If you have done a comprehensive patient interview and physical exam, you will have a good baseline with which you can compare your results. Outcomes include both objective and subjective elements as well as both biomedical and acupuncture components. A lack of objective measures does not necessarily confirm a lack of effect, although they certainly will make the case more credible. Ideally, multiple gauges of outcome may be used to compare the patient's condition before and after a treatment course. These include biomarkers as a tool of measurement, Chinese medicine diagnostic findings, any of the outcome factors mentioned in the previous chapter, and the patient's subjective experience. Observing outcomes from multiple perspectives may help increase the validity of your case report.

As is the case with diagnoses, some caution must be exercised regarding biomedical lab values. Intrinsically, treating a patient with acupuncture for the sole purpose of altering a lab value may not necessarily help the individual. For example, if the goal of acupuncture treatment is to lower blood pressure in a patient, this may or may not actually benefit the outcomes of the patient (similar to the questionable effect of atenolol medication on overall mortality as described in chapter 3). If the outcome factors of the patient are improved by observation, including lab value improvement may be an added bonus for the sake of integrative discussion with biomedical practitioners, but improving measures of function are more important than reducing measures of risk.

Subjective outcomes relevant to the acupuncture paradigm may also add to the depth of your case. Reporting any significant changes in the tongue, pulse or other palpatory findings for example may be helpful as a way to ascertain if a traditionally diagnosed pattern has changed. Noting any improvements besides the chief complaint that might have been influenced positively by the treatment of a pattern can support the use of traditional acupuncture as a holistic therapy. Citing the patient's subjective experience of acupuncture and its benefits may also be helpful to report, especially in cases involving mental-emotional complaints (as these elements are best captured in the patient's own words). Some sources encourage directly quoting the patient to describe subjective outcomes and to verify that the patient's needs, preferences, and concerns have been met in the context of evidence-based medicine.[5]

Complex treatment courses and outcomes

For a number of reasons, in some complex cases, your treatment might shift midway through the treatment course. For this kind of case, presenting the treatment chronologically in a number of phases may be easier for the reader to follow. For example, one case where there may be a shift in the treatment could involve an initial incorrect diagnosis and treatment with poor results, but after the diagnosis was changed and with subsequent treatment, the patient responded well. Another type of case that involves multiple phases is when the first treatment style or approach resulted in slow progress, but shifting to a second treatment style or approach resulted in a more significant difference in the patient's response. Still another might be that the first phase of treatment was designed to address the branch problem (acute symptoms) with the second phase addressing the root (chronic underlying pattern of disharmony). Another might be that the first phase addresses one area of imbalance (e.g. excess) before attacking a secondary pattern (deficiency).

When writing a case with multiple phases, each phase should be described at the same level of detail as in simple cases. After the initial intake is described and diagnosis is made, the initial treatment section should open with a statement of the overarching treatment course. For example, you might begin with 'The patient received a total of ten weekly treatments, in two phases. Phase 1, which included treatments 1-4, was focused on calming the acute flareup of abdominal pain. Phase 2, which included treatments 5-10, was focused on nourishing the Spleen and Stomach to resolve the underlying pattern that made the patient susceptible to acute flareups.'

Then proceed to describe the treatments for phase 1, followed by outcomes for phase 1. Describing outcomes between phases is crucial because it allows you to explain to the reader why you chose to move on to a second phase. Phase 1 outcomes should be followed by a transition to explain the reason for changing the treatment plan. For example, you might observe that 'After taking herbs for two weeks the patient's pulse had become less wiry. Because it appeared that the primary excess pattern had resolved, the treatment goal was shifted to nourishing his underlying deficiency,' or 'Because the patient's symptom had not seen any positive change by the third treatment, the treatment style was changed from a Japanese Meridian Therapy approach to the Balance Method style.' Once the reasons for the transition have been stated, the patient's second phase of treatment can then be described in a fashion similar to the first phase (treatment principles, point prescriptions with rationale, details on treatment specifics, and possibly a table for clarity). After the second phase, outcomes can be described. If there was a third phase, it will follow in the same manner.

Especially in a complex case, the reader has to understand why you made the treatment decisions you did. The case report is not just a means of presenting an effective point prescription for your patient; it is also a way to describe to your reader how to navigate a potentially complex case.

Timeline

Although this particular aspect of case reporting has not been widely adopted at this time, a timeline contained within the case description section may help to clarify the sequence of relevant events for complex cases. A timeline provides a chronological visual representation of key events in the patient's experience. It covers relevant elements of the patient's health history, as well as indicating the timing of therapy and outcome measurements. In the case description section, a timeline might include key events that may have given rise to the patient's condition, the biomedical diagnosis, periods of biomedical treatment, and traditional acupuncture treatment regimens. The timing of any relevant lifestyle changes, dietary modifications, work stresses, or adjunctive treatments may also be included.[5] The timeline can help you to organize your writing especially in cases with a complex patient history or treatment course. In some published case reports, instead of using a graphic timeline, tables can be used to clarify the timing of events.

Prognosis

Including the patient's prognosis in a case report is optional. Some journals do not include prognosis because it is based not on facts, but on experience and prediction. If you choose to include a prognosis, its scope will depend on what guidance will be helpful for your reader. How do you expect the patient to fare after treatment ends? Do you anticipate that the patient's symptoms will return, and do you anticipate the need to continue treatment at regular intervals? If so, for how much longer? What instructions have you given for follow-up? How will you know when you will be able to discontinue treatments? For answers to any of the above questions, you must provide a reason for your assertion. At the end of this section remember to avoid any commentary and save it for the discussion — the case description focuses on the facts of the case.

CHAPTER 9

Writing the Introduction Section

THE INTRODUCTION SECTION of the case report, also known as the background section, provides background information on the condition being treated.[1] It prepares readers by reviewing existing knowledge relevant to the case. The introduction should be a review of the most recent information available on the topic from reliable sources. For acupuncture case reports, the introduction material often comes from two different perspectives: biomedical sources and traditional acupuncture sources. The main goal of writing this section is to provide enough background material so the reader will not have to search for additional information in order to understand the case description and the points made in the discussion section.

It is recommended that you write your case description section first (described in the previous chapter) so that the topic and scope of your introduction section is clear. The introduction usually contains three content areas: biomedical background material, acupuncture background material, and a research review (do not use separate headings). The introduction section should have no information about the patient in your case.

Framing the Introduction

Writing the introduction can be challenging because readers may vary in knowledge and experience. When planning this section, consider your audience. If you are writing to publish in a biomedical journal whose audience tends to be less familiar with traditional theory, more detail can be included about the biomedical perspective, while possibly reducing the acupuncture background material. If you are writing to publish in an acupuncture journal, more detail can be included about the traditional perspective to encourage deeper discussion, while possibly reducing the biomedical background material.

Your readers may range in experience from beginning students to seasoned practitioners. Information you choose to provide in this section could be too elementary for advanced practitioners, while novice acupuncturists might desire more material. The level of information depends on the knowledge base of the reader. If you are writing a case report as part of academic coursework, imagine your audience as recent graduates of an entry-level acupuncture program with practitioner-level knowledge of both biomedicine and acupuncture theory. The instructor should be able to provide a few sample academic case reports to illustrate the expectations for the content of this section. When writing for a journal, you will find it helpful to read sample case reports from previous issues to understand what amount of background information is appropriate, and where to focus your content. Some journals focus on the biomedical material, some more on the traditional acupuncture material, while some fall in between by balancing the two perspectives.

A common error is to pitch the level of language to a layperson when intending to submit an article to a professional journal read by trained practitioners. Just as an article on infertility written for a biomedical journal should not contain definitions of estrogen and progesterone, likewise an article written for an acupuncture journal should not contain the basic definition of *qi*. Thought should also be given to your audience and the language you use for the introduction. As a result of all these factors, the length of the introduction can vary greatly depending on the topic and the intended audience.

Biomedical Introduction

The biomedical review of knowledge relevant to the condition should include a definition of the condition, demographics, understanding of mechanism/

cause, symptoms experienced, commonly used biomedical treatments, efficacy/risks of treatment, and prognosis (generally presented in this order).

One of the first tasks in constructing a biomedical introduction section is to determine the topic. This topic is the condition that the patient has been treated for. If the patient has one major condition being addressed, the choice is straightforward. For example, if a patient was treated for adhesive capsulitis of the shoulder and all treatments were designed to address increasing shoulder range of motion and decreasing pain, the biomedical introduction section should focus on that. Some cases involve a patient where multiple related conditions are treated simultaneously. If multiple conditions were treated, the focus should be on the condition primarily addressed by the treatment. For example, if treatment in the case focused on low back pain, but the patient also happened to receive occasional treatment for allergic rhinitis, low back pain will be the topic, and allergic rhinitis will likely not need to be covered. For a patient treated for infertility and endometriosis, if the primary goal was to reduce the heavy and painful periods, endometriosis might be more appropriate as the focus, with an added short paragraph on the effects of endometriosis on fertility.

In some cases, two unrelated conditions are simultaneously addressed more or less equally. Consider a case report where the patient was treated for both peripheral neuropathy and vertigo. The acupuncture diagnosis revealed that both conditions had the same underlying pattern differentiation, and when treated as such, both conditions improved. As the two symptoms were treated equally with the same set of acupuncture points, the biomedical introduction would have a section on peripheral neuropathy, plus a second section on vertigo (although both should be pared down so that the section is not exceedingly long). If a case happens to have two conditions commonly seen together, such as diabetes and hypertension, you may find a diagnosis that includes both, such as metabolic syndrome, or cover each in turn and add an additional paragraph that shows the relationship between these two conditions.

Some cases involve a patient who does not have a specific biomedical diagnosis but instead presents with a mystery symptom not linked to a known mechanism or cause. For this type of case, the introduction could be used to elaborate on the symptom itself and its biomedical differentiation. In a case where the patient is being treated for extreme fatigue, for example, it would be useful to differentiate and describe the variety of possible biomedical causes for that symptom, followed by a brief review of treatments for each of the causes. If a symptom is an adverse effect of biomedical treatment, it may be appropriate to add an extra paragraph describing the symptom in that context.

In very rare circumstances, the topic of the introduction may be derived from significant discussion points of your case. For example, if your patient suffered from chronic ankle pain which did not respond until his emotional trauma was first treated, it would make more sense to choose emotional trauma as the topic, as it would be more relevant to explore than differentiating and discussing the causes and types of ankle pain.

Once the topic of your biomedical introduction has been identified, your next task is to compile information from a variety of recently published biomedical sources (ideally published within the last ten years). The *definition* of the condition should be followed by details on the *diagnosis* of the condition, whether it might be specific lab values, or perhaps a list of multiple criteria for a biomedical syndrome diagnosis (e.g., Irritable Bowel Syndrome). Then describe the *symptoms* typical of the condition. *Demographics* regarding the condition should also be stated if available, including its prevalence and whether it is more likely to occur for a certain gender or age or ethnic group.

Standard *treatment(s)* should also be described, along with circumstances when they are typically used. List the classes of medications routinely prescribed (including generic names of individual medications and their common side effects and risks). If applicable, surgical options should be discussed. Limited complementary and alternative therapies may be briefly mentioned if they are common in practice (including nutritional and herbal supplementation). If the patient in your case has specific procedures, medications, or reactions that are of particular interest to your case, this material may also be useful to include.

Finally, the typical *prognosis* of individuals with the condition should be described. With all conditions, you will be able to say how well the typical treatments help. With some, you will be able to additionally provide information as to how patients with this condition tend to fare without treatment. Doing so will provide a reference point for the discussion section later on, where you will relate the outcomes of the specific patient in your case compared to a typical outcome, which may contribute to the potential significance of your case.

This section should be written in the present tense to indicate that the information obtained from the medical literature is current. The only exception would be information that describes a historical approach to diagnosis and treatment, which may be included if appropriate.

Biomedical References

When surveying the current knowledge of the condition and its treatment, the biomedical introduction material should be cited from a variety of sources. The most respected sources of information for this section are peer-reviewed articles. These may include recent articles on the topic of diagnosis and treatment options for the condition being described. Systematic reviews of primary research (e.g., Cochrane reviews) are some of the most well-accepted sources on the topic of diagnosis and treatment. Recently published papers on practice guidelines may also be useful. Articles on the diagnosis of a condition can be cited, and individual clinical trials testing the efficacy of specific treatments ought to be included. Citing sources written by acupuncture experts for biomedical content is not advised, as they are not considered experts in biomedicine.

Searching a biomedical database such as PubMed will generally provide the best and most current sources. It is preferable that you cite the material directly from peer-reviewed studies, systematic reviews and clinical trials. Recently published editions of biomedical textbooks are not quite as strong as peer-reviewed articles; if you need to resort to a textbook, choose one with a specific focus (e.g., a text on endometriosis is preferable to a text on gynecology if endometriosis is the topic of your case). If you choose to reference online material, it should come from a source that caters to medical professionals (e.g., the website of the

American College of Rheumatology). However, as the material found on such websites is often derived from peer-reviewed primary sources, it would be much better to cite from the original papers cited on the website. Avoid citing paid subscription websites like www.uptodate.com, as a reader who does not have paid access to these sites will not be able to use your references to read further on the topic. (Also, the information from these reference websites is compiled from original peer-reviewed research in the first place). Finally, avoid sources written for the layperson. Websites written for the general public are not considered scholarly resources and should not be used as references (e.g., WebMD or the Mayo Clinic). Databases such as EBSCOhost and PubMed can be useful for as you search for professional sources. While searching Google is not advised, searching Google Scholar may turn up worthy academic articles, but is not as thorough as using biomedical databases.

Plagiarism is the act of unethically taking credit for someone else's work. Drawing on the ideas of others to support your claims is an important aspect of research; however, when doing this, it is crucial to give credit to the original author. Plagiarism comes in three main forms: 1) copying text from a source without quoting and including a citation, 2) paraphrasing or summarizing without referencing the source, and 3) using the ideas of others and claiming that they are your own, without acknowledging the source;[2] if you quote a passage in your draft, make sure you include quotation marks. Try to reword or paraphrase source material; if you need to quote a passage in your draft, make sure you include quotation marks. In order to avoid plagiarism, make sure any ideas which originated from another source have citations, whether paraphrased or in the form of a direct quotation. The format of the citations should be consistent with your references section, and *Instructions for Authors* should be followed carefully.

Acupuncture Introduction

The goal of the traditional acupuncture portion of the introduction is to provide background material from a variety of sources that will make the case easier to comprehend. This usually consists of a review of the condition from a traditional perspective, an introduction to the techniques and approaches used in the case, or sometimes both.

If you are writing your case report for an academic assignment in an acupuncture program, or if your aim is to publish in an English language journal that focuses on traditional acupuncture, your reader will likely have a basic professional knowledge of theory. The acupuncture introduction should be written so it is not too elementary (for example, including information on the basic nature of *yin* and *yang* is not necessary). A fair assumption for the minimum level of reader knowledge is material that typically is covered in entry-level acupuncture degree programs. One of the challenging aspects of writing this portion of the introduction is that acupuncture colleges often cover a variety of approaches within the curricula, although not all approaches are taught at each college. Because a large percentage of acupuncture colleges cover TCM acupuncture as part of the basic curriculum, TCM is often used as a reasonable frame of reference for case reports.

The content of the acupuncture portion of the introduction will depend upon the content of the case. If the case involves the TCM approach to diagnosis and treatment, the content of this section will begin with identifying the disease diagnosis. The disease diagnosis is generally a two-syllable Chinese word (sometimes three). Examples of disease diagnoses include *bi* Syndrome (*bi zheng*), headache (*tou tong*), epigastric pain (*wei tong*), infertility (*bi yun*). Most acupuncture colleges teach TCM pathology courses according to these topics. After defining the disease diagnosis, provide the pattern differentiation and list the TCM patterns most likely to occur in patients with this disease diagnosis. Each source text may have a slightly different wording about the possible patterns involved, so be sure to combine the information from all of your sources and include all of the basic patterns without overlap. In some journals, TCM patterns and their related information are presented in chart form, including typical patterns, key signs and symptoms, and sample acupuncture points for each pattern. If your case patient was prescribed Chinese herbal medicine, you will need to include some additional background information on herbal formulas that might be used for the pattern or condition.[xi] If you decide to use a chart, include any other special considerations for treatment in paragraph form after the chart. To save space, a written description of the patterns and treatments may be included instead.

In case reports describing a patient treated with a non-TCM approach to acupuncture, you will need to compile source material relevant to your case. Briefly review basic ideas about the approach you used, since not all schools offer training or introduction to specialized approaches to acupuncture and Chinese herbal medicine. Some of this material may involve the classics. For example, if your case employs the *Shang Han Lun* theoretical approach, it may be useful to cite and explain relevant passages from that classical text. As preparation for the case description, a beginner to any non-TCM approach would benefit from an introductory overview of how diagnosis is performed and how treatment is generated from that approach. As with all sections, the level of detail will depend on the journal you intend to submit to.

Some journals place the information about a non-standard TCM approach in the discussion section rather than in the introduction. This is not ideal because uninitiated readers may be confused while reading through the details of the case. Most would find it useful to read general information about the approach in the introduction before reading about its specific application. After providing a brief introduction to the specialized style of acupuncture used in your case, you will want to discuss any additional considerations that may help frame the case and make it easier to understand.

Acupuncture References

A variety of sources should be referenced here, including but not limited to general Chinese medicine texts, classical texts, journal articles, and modern texts that focus on the condition discussed in the case. A summary of the relevant chapter of a general TCM medicine text is not

---

xi See Appendix D on Chinese herbal medicine.

adequate for the acupuncture portion of the introduction. Articles published in journals catering to Chinese medicine practitioners may be useful to cite, as they often focus on particular conditions and contain more specialized information than what is commonly found in general texts. Unfortunately, most of the journals with articles on the traditional practice of acupuncture are not included in online searchable databases such as PubMed, and each journal may require an individual membership to access these articles. If you only use biomedical databases to search for your background material, this can mean you may miss a significant portion of the acupuncture literature. One useful resource is the library at a local acupuncture college, which may have either paper copies or online access to these articles. Textbooks and articles describing non-TCM styles of acupuncture may be used as reference, but lectures and personal correspondence are acceptable only if the material is not available anywhere else in written form. Acupuncture material written for layperson audiences (including but not limited to practitioner websites and acupuncture newspapers) are not considered academic sources and therefore should not be used as references. If you are uncertain about the proper translation of any Chinese medicine terms, a useful reference is Wiseman's *Practical Dictionary of Chinese Medicine*.[3] If you feel that the particular wording of the source is important to preserve (for example, if you are quoting a classical Chinese text), place the source (or translation) in quotation marks and cite. For quotes longer than two lines, the passage will need to be single-spaced and indented without quotation marks. For classical Chinese texts which have chapter numbers (e.g., *Huang Di Nei Jing Ling Shu* Spiritual Pivot), including the chapter number with the quote is helpful to the reader. For any direct quotes from classical or modern texts, be prepared to provide page numbers.

Research Review

After the theoretical review of traditional diagnosis and treatment, the scientific research is then reviewed. This most often is presented as a review of acupuncture's efficacy in treating the condition being discussed. This review should be fairly brief and contain information relevant to the case. The goal is to provide a context that can allow the reader to understand how your case contributes to the existing literature. A full literature review is often considered an article in itself; combining a full literature review with a case report would be too lengthy and the paper could lose its focus.[4]

Searches on PubMed should be a part of your process when preparing to write this section. Readers will want to know what evidence currently exists for how well acupuncture treats the condition being discussed. If Chinese herbal treatment is central to the case, it is important to also include herbal research. Sources for this section include systematic reviews, individual clinical trials, pilot studies, and case reports. Systematic reviews (or meta-reviews) are articles written by researchers who examine the quality of the conclusions of multiple clinical trials to consolidate results and thereby make stronger generalized conclusions. You may also choose to review one or more individual clinical trials. Pilot studies may also be reasonable to include if clinical trials are not available or if a particular pilot study has relevance to your case. Case reports themselves can be reviewed on the specific topic — collectively or individually — to provide an opportunity to enrich the discussion section.

For each piece of evidence that you include, briefly state the nature of the study and the results as well as information about the protocol if relevant. While this type of information might feel a bit more biomedical in nature, it belongs at the end of the acupuncture introduction, since some of the points you wish to emphasize about the trials may refer to traditional theory. If your condition is quite unusual and no clinical trials have been performed, you may instead include efficacy studies on a very closely related condition. For example, if your case is on psoriatic arthritis but you are unable to find any acupuncture trials on this condition, you may cite articles on a similar condition such as rheumatoid arthritis (with a sentence specifying how the condition is related). If an online database search for your condition reveals no results, you do not need to explicitly say so. Your readers are much more interested in what you have found rather than what you haven't found.

Lastly, if you wish, the only acceptable mention of the patient in the introduction is an optional transition statement to close out the section. For example, the following would be an acceptable conclusion to the introduction section: "The purpose of this report is to describe a case of biopsy and MRI-confirmed ductal carcinoma in situ treated with acupuncture and Traditional Chinese Medicine that regressed over a period of 3 years."[5]

While the introduction section may appear to add little to the overall paper, it can be very helpful to prepare the reader to fully understand the significance of your case report. For acupuncture students, introductions are often used as an exercise to demonstrate a thorough understanding of both biomedical and traditional diagnosis and treatment. While this is a worthwhile goal for academic programs, the introduction to a published article must more flexibly cover the background information most useful for understanding the case. Depending on journal readership, introductions may focus more on the biomedical perspective or more on traditional acupuncture. Writing to serve the needs of your target audience is the key to a useful introduction.

CHAPTER 10

Discussion and Conclusion

THE DISCUSSION SECTION is your opportunity to reflect on the significance of the case by explaining what is original in your work and why your results are important. The discussion section puts the case into context, explains what happened, and explores the implications.[1] As such, the discussion is arguably the most difficult section of the case report to write.

The ideas that inspired you to select the particular patient for your case report may also help you to select the discussion topics. It may therefore be worthwhile to revisit the reasons why you initially chose the case to write about. Choosing a case merely because the results are good may result in a flat discussion. A discussion section should not simply repeat the results but must extend beyond to explain the significance of the case.[2] At the same time, be sure that your main points are not too broad-reaching and do not generalize too widely. Many papers are rejected by editors because of a poor discussion section if it does not present anything new or if the recommendations are not supported. In the end, preliminary acceptance will depend on the report's contributions to current knowledge about the condition.

The exact contents of the discussion section vary widely from journal to journal, and while a number of possible topics and directions have been taken in past case reports, there is no standard organizational format or content list. That being said, the discussion section typically covers several general topics: a summary of the case, observations made, and recommendations for the reader.[xii] As you cover these three areas of discussion, keep in mind that case reports are limited as far as assertions that can be made. A case report cannot prove that the treatment described is actually what caused the result. Any number of variables may cause or contribute to outcomes, such as other concurrently used modalities or lifestyle changes that the patient may or may not have told you about. Alternatively, the outcomes may be due to spontaneous recovery, although this may be more or less likely in your specific case. Nor can a case report prove that the experience of just one unique patient is typical. Do not go beyond the evidence and draw unjustified conclusions about the generalizability of your one case.[3]

Summary of the Case

The first task in the discussion section is to summarize the case and to describe how it relates to previously published work. To provide a transition from the case description section, many writers begin this section with a short 2-3 sentence summary of the important features of the case (including the patient presentation, diagnosis, treatment, and outcomes). The reader has likely just finished reading the case description, so there is no need to go into too much detail. For example, consider this summary of a case of ductal carcinoma in situ:

> This report describes a case of ductal carcinoma in situ (DCIS) in a perimenopausal woman, treated by excisional biopsy followed only by CAM care without the use of chemotherapeutic agents or further surgery. The patient experienced a 75% reduction in volume of her DCIS between diagnosis in 2009 and November 2011, concomitant with the use of alternative therapies.[4]

After this brief synopsis, it is customary to include a description of the strengths and weaknesses of your case report. When reporting strengths,

---

xii Within the actual paper writing the subheadings 'Summary', 'Observations' and 'Recommendations' is not necessary.

be careful to avoid exaggeration. Also include any reasons relevant to your case that may explain why the results might not have the power that they appear to. While weaknesses might seem to weaken your case at first, being realistic will make your case more unbiased. The following are example statements describing some possible strengths and weaknesses of a case.

Example strengths could include one or more of the following:

i. 'Because the patient in this case had been experiencing the condition for over ten years without improvement, and the relief began shortly after starting acupuncture treatment, it is likely that the acupuncture treatment played a part in his healing process.'

ii. 'Even though this chronic condition is characterized by periods of aggravation and amelioration, the number and severity of flareups experienced by the patient decreased significantly for over a year after a ten-week course of acupuncture, which is unusual.'

iii. 'The patient received multiple modalities of Chinese medicine for the first several weeks of treatment, but the improvements seemed to markedly improve after the fourth treatment when the additional modality of bleeding therapy was applied.'

iv. 'The patient reported that she did not begin any other alternative therapies or significant lifestyle changes during this time.'

Example weaknesses could include one or more of these:

i. 'Because this particular condition has a tendency to resolve spontaneously, it is difficult to ascertain whether the patient's symptoms would have improved on their own within this time period.'

ii. 'Because this chronic condition by nature is characterized by periods of aggravation and amelioration, it is possible that this patient may have improved without treatment.'

iii. 'Because multiple therapies were applied concurrently, it is difficult to say which of these caused the patient's improvement.'

iv. 'Because the main outcome measure was a subjective verbal pain scale, the results of the case are patient-reported outcomes and not objective validated measures.'

Finally, a statement comparing the case to relevant published literature will round out the summary. This comparison may be in relation to acupuncture clinical trials or published case reports.[5] Your case may uphold published research study findings, challenge them, or differ from previously published case reports in one or more respects.

Consider the following range of examples of statements pertaining to significance:

i. 'In all of the clinical trials in the published literature, the style of individualized TCM acupuncture was applied. In this case report, the use of the Balance Method is unusual and deserves further examination, as no shoulder pain clinical trials have been published that examine this style.'

ii. 'In the successful 2007 GERAC acupuncture clinical trial for chronic lower back pain, patients with previous spinal fractures were excluded.[6] Since the patient in this case report had a history of spinal fracture, it is worthwhile to take a closer look at this particular case to see how one might modify treatment to address this specific issue.'

iii. 'In the only published clinical trial treating this condition, the successful protocol involved sessions twice per week over a period of five weeks. Because the patient in this case report received complete relief within three treatments, it is worth examining what factors may have contributed to her quick response.'

iv. 'There are currently 13 published individual case reports accessible through the PubMed database which describe an acupuncture intervention treating migraine headache, but none describe a case where the migraine headache was associated with food sensitivity.'

v. 'This case reports a successful outcome for a condition that has little literature. As it is a rare condition, no acupuncture clinical trials have been completed and no case studies have been written.'

Observations

After a brief, succinct review of the case, the discussion should describe what makes your case distinctive. Because each case is unique, there are no standard steps to adhere to in a discussion. The nature of the case and the level of your audience will define what you develop as the significant points. The following is a list of 'discussion seeds' for possible points that you may wish to discuss, depending on if they apply to your case. For any point you choose to discuss, you'll need to expand on the seed sentence in a paragraph so that it fully explains your idea and puts it into context.

Explain what makes your case unique

After comparing your case to the research literature, one possible next direction is to delve into the specific aspects that make it unique. For example, imagine that you have framed a lower back pain case as unique because no trials have tested Balance Method treatment for back pain. Logically, it might follow to describe a bit more why the Balance Method style of acupuncture may have been a more useful choice for this case. More about this style of acupuncture and its strengths can be described and explained in the discussion. If your case involves a style other than TCM acupuncture, you may wish to consider describing in more detail how you arrived at your diagnosis.

Perhaps your case is more complex than typical, and due to that complexity evidence from clinical trials might not apply as readily. For example, patients with lower back pain as well as a history of spinal fracture may have been excluded from most clinical trials. Given this, you can describe how your treatment strategy was modified. Or for a patient who recovered much more quickly than expected when compared to other case reports or clinical trials, you may then go on to discuss what made your treatment or this patient unique. Cases where unexpected benefits were realized after treatment also apply here, as do cases describing successful treatment of recalcitrant conditions.

Explain how your case demonstrates a specific theoretical concept

Classical Chinese medicine texts can be difficult to translate and difficult for readers to understand depending on their degree of familiarity with the acupuncture paradigm. If your case highlights an interpretation that is supported by successful practical application, this would be a worthwhile discussion focus that can help guide other practitioners with similar cases. Biomedical journals may be less likely to attract readers interested in traditional theory, but acupuncture journal readers may be eager to learn under what circumstances a theoretical statement may be applied to generate an effective treatment for an individual patient. Demonstrating how to make an abstract theory clinically useful, even if it is simple and straightforward, helps to grow the collective wisdom of practitioners. For many beginning case writers, choosing a case that demonstrates a successful application of a specific acupuncture theory may be a good way to get started writing case reports. If you have had numerous patients getting similar results with the theoretical interpretation suggested in your case, it is acceptable to mention what you have found to be a typical response, along with your level of experience.

Discuss any special considerations in making the diagnosis

Part of the reason for writing a case report is to guide a less experienced reader through the process of diagnosis. Making a diagnosis for a given case may be challenging for a number of reasons. You may wish to point out to the reader the key observations that were instrumental in your diagnosis. Perhaps a specific factor determined an effective direction for your treatment. Your case may demonstrate an unusual pattern for a condition, such as Lung *qi* deficiency leading to headaches. A complex case might involve multiple patterns, such that the order in which you chose to address the patterns was crucial for resolving the case. Perhaps different symptoms arose after successfully addressing the first pattern, and a new diagnosis was called for.

Alternatively, perhaps your case involves an initially incorrect diagnosis that led to ineffective treatment, but when the diagnosis was revised the subsequent treatment yielded better results. In this type of case, discussing in retrospect why the original diagnosis was incorrect and what specifically led you to revise your diagnosis will be helpful to the reader.

Discuss any special aspects of the treatment or treatment approach

If there are features of the treatment that require additional explanation, you might want to clarify how and why those features are appropriate for your case. For example, you may wish to elaborate as to why a specific point or point combination was a particularly apt treatment. You might want to discuss its history/background or the traditional theory behind why it may have been useful in this situation. Some practitioners add points that have been shown to have specific biomedical mechanisms; if this also influenced your point choice, provide a citation. Or, if you have used a special technique or new technology that is not in common practice, you can describe the procedure more in detail. If your management of the case is unusual, this might also be discussed (for example, this may be a decision to shift direction or change the treatment frequency halfway through the treatment course). If you wish to elaborate on specific aspects of the treatment, take care that you don't simply repeat a description of what happened.

Carefully explain a biomedical mechanism

After comparing the case to the literature, some writers choose to discuss biomedical mechanisms of acupuncture. This tends to happen more often in biomedical journals where the audience is more likely interested in mechanism and possibly less likely to have a background in the traditional theories of acupuncture. Ideally, the mechanisms discussed in case reports are well-documented, and citations can point the reader back to the source material. For example, consider this discussion of the mechanisms behind *guasha*:

> According to the literature, several mechanisms of *guasha* may explain the immediate short-term reduction in shoulder pain experienced with each treatment. *Guasha* has been shown to increase local microcirculation by 400% in the first 7.5 minutes with a continued increase for 25 minutes post-treatment. Another study has shown an increase in tissue temperature.[7]

If you want to draw correlations between biomedical mechanism and traditional acupuncture theory, it is important to make sure that you clearly state that what you are proposing is a possible correlation that has not yet been proven. Avoid any phrasing that may cause a reader to assume that the connections are well-accepted. As discussed in the earlier chapter on integrative medicine, a thought exercise on the overlap of biomedical mechanism and acupuncture theory may bring up some interesting points, and there may be instances when this type of translation leads to new ideas that may be effective in practice. In any new recasting of theory, the ultimate test is the response of the patient. But simply defining acupuncture concepts in biomedical language may lead to incorrect assumptions about traditional theory, which in turn might lead to new ideas that may not be effective in practice.

Because Chinese medicine employs some of the same words as biomedicine, it is easy to draw careless connections that may inadvertently lead to confusion. For example, asserting from one case that patients with abnormal liver enzymes are likely to have Liver *qi* stagnation is a logical error. And as discussed in chapter 5, characterizing all conditions ending in '*-itis*' as Heat in the TCM context because of the etymology of the word 'inflammation' may lead to ineffective or misguided treatment. Whether or not the correlation you make between biomedicine and acupuncture is put forth as useful or questionable, be very cautious and transparent to the reader when making these connections.

Here are a few suggestions to guard against this type of misconception based on a false equivalence:

i. Consider using the words 'may be' instead of 'is.' Using 'in some cases' may also qualify your statements. For example, 'Taking certain types of antibiotics in some cases may have a similar effect to the Chinese medicine concept of introducing Cold into the Stomach.' Or, confine your hypothesis to the patient: 'For the patient in this case, it appears that the reduction of *yin* deficient symptoms was coincident with a rise in estrogen.'

ii. Include a disclaimer: 'This may not be true for all cases; performing traditional diagnosis will determine if a patient falls under this pattern and therefore may benefit from the type of treatment provided in this case report' or 'Not all patients with Kidney patterns will necessarily show an imbalance in adrenal hormone levels, and hormone levels are not used traditionally to determine TCM treatment.'

iii. Preface the traditional acupuncture term with 'Chinese medicine' and the biomedical term with 'biomedicine' and/or using capital letters for Chinese medicine terms and lowercase for biomedicine: 'Introducing a biomedical anticoagulant medication is not the only way to achieve the Chinese medicine concept of moving Blood stasis.'

iv. Keep the biomedical perspective and AOM perspectives in separate sentences: 'In this case, a patient with chronic severe eczema was treated successfully by addressing his longstanding Kidney *yang* deficiency pattern. It is interesting to note that in biomedicine, prednisone is a medication often used to treat chronic severe eczema, and prednisone is a medication that has a strong effect on the adrenal glands.'

Explain the use of integrative medicine in your case

If your case involves multiple modalities beyond acupuncture, another possibility for a discussion topic is how the modalities work together from a practical point of view. You may wish to discuss any of the following topics:

i. Coordinating with a practitioner of another modality. For example, as you treated a patient for lower back pain, while at the same time his biomedical physician was weaning him from opioids, describing the interaction and the process would be helpful for readers with a similar patient.

ii. Providing acupuncture to a patient in preparation for a biomedical procedure, or offering it after a biomedical procedure to speed up recovery and reduce adverse effects. Your discussion may briefly describe the biomedical procedure as well as how acupuncture could potentially help. Prognosis details also may be useful here (e.g., patients recovering from hip replacement surgery recover in

10-12 weeks on average, but your patient was able to resume most activities after five weeks with acupuncture).

iii.   Combining any two modalities in a thoughtful way. While many would consider integrative medicine to be the use of an alternative therapy alongside biomedical therapy, it is possible to integrate any two therapies. Remember that the sophistication of integrative discussion may be expanded by considering juxtaposing the strengths and weaknesses of the respective sides.

Comparing two or more cases

If you are presenting two or more cases, comparing them as a case series will allow for a greater depth of explanation. If the reader is looking to understand how to diagnose and treat a patient with a similar condition, having not only one but two examples to work with can be useful (see the following example), especially when you offer additional guidance on diagnosis or how to refine treatment choices to match the individual patient:

> The intermittent nature of case one indicates more qi than Blood stagnation and demonstrates treatment of an early presentation of plantar fasciitis. Case two is a typical presentation of plantar fasciitis where Blood stagnation dominates.[8]

If you are contrasting two or more cases, a table may be helpful to tease out the differences between the individuals and how their treatments diverged. Writing a case series is more involved than a single case. If you are writing a case series for an academic assignment, assume that the writing process will be longer and more involved than for a single case. If you are writing to publish, the case descriptions will likely need to be abbreviated in order to fit the material within the fixed word count dictated by the journal.

Recommendations for the Reader

After taking the reader through the significant observations of your case, the third and final portion of the discussion section should contain relevant recommendations for the reader. Any recommendations you make should be supported by the observations in your case, and any reasoning behind your recommendations should be logically described. Recommendations may have ramifications in terms of changing clinical practice, education, or research. This is the ultimate statement of significance for your case.

Because you are writing about an individual, remember to temper your recommendations accordingly. For example, even if your patient has a remarkably positive response to acupuncture to manage his diabetes, it wouldn't be appropriate to phrase your recommendations in a way to state that 'all' patients 'should' receive acupuncture for this condition (although it is acceptable to use 'should' in a cautionary context, e.g., practitioners should be careful with needling depth at the acupuncture point LU 1 (*Zhongfu*) to avoid causing pneumothorax). The strength of any recommendation needs to reflect the nature of the case, and using words like 'may' or 'could' softens the phrasing to make the recommendation align with what can be reasonably derived from the actual case.

Recommendations pertaining to clinical practice

The nature of clinical practice recommendations can vary widely from case to case. The simplest type involves suggesting that acupuncture treatment be considered when encountering patients with similar situations. Obviously, this type of statement might not be worthwhile when the topic is a condition already commonly treated in acupuncture. This type of recommendation may be more appropriate for a lesser-known disease or for a recovery from an unusual condition that has not been thoroughly researched.

A second type of clinical practice recommendation involves giving specific suggestions aimed at improving patient care. You may advise your reader on what to consider when making a diagnosis or on what types of styles or treatment techniques might be useful for a similar patient. You may advocate trying a new or unusual technique or using a particular interpretation of theory as long as these are directly relevant to your case. For case reports describing adverse events, advise your reader on how not to repeat the mistake made in the case. Do this by raising red flags that specify which points and techniques require special care and why, and by instructing on proper follow-up procedure if should this adverse effect arises in clinic.

Making recommendations regarding management of the case are appropriate here as well; perhaps the dosage of acupuncture (i.e., number of treatments) may have made a significant difference in the outcome of your case. If your case describes an integrative treatment strategy, for similar cases you may recommend for similar cases that acupuncturists connect with their patients' primary care physicians or other providers to explore collaboration.

Because every patient is unique and the case report describes a single patient, it is unlikely that the patient of a practitioner who reads your case would be a perfect match. As such, your reader will need to reconsider whether any recommendations you make are appropriate for each patient, and if so, to potentially modify the treatment to fit his or her situation. With this in mind, it may be useful to indicate from your experience what types of patients you believe these recommendations would be best applied to. This might involve extrapolating to a larger group. For example, if your case patient recovers successfully from migraine headaches caused by gluten sensitivities, you could suggest that in theory this type of treatment might be considered for migraine caused by other food sensitivities. Alternatively, you may cast a smaller net and exclude inappropriate patients. In this instance, if your case patient recovers successfully from migraine headache caused by gluten sensitivities, you may indicate that your treatment strategies may not necessarily be as useful for treating patients with migraine characterized by other causes or patterns. If you believe that your treatment may be more useful for one patient population versus another, you may also discuss it here as long as it is backed up by your case.

If you wish to recommend acupuncture for a condition where efficacy studies have generally been negative, you will need to explain why you support the recommendation despite the literature (e.g., perhaps the research studies did not test the acupuncture style you used or your treatment frequency was increased beyond that of research testing methods in the past.)

Recommendations pertaining to acupuncture research

Many published case reports conclude with a statement recommending that 'further clinical trials need to be done' to study the efficacy of acupuncture for the condition at hand. However, while a case report may help us to generate a hypothesis that may be investigated, it is unlikely a case report will trigger the significant energy and funding required to carry out a clinical trial. Recommending that more research be done is a bit of a throwaway, especially if this type of statement is thoughtlessly tacked onto a paper describing a condition that has already been studied extensively.

In the context of scientific research, a case report describing a patient intervention is not merely a single data point indicating success or failure. Consider instead how details of the case may be able to inform how research study design can be improved. For example, if your case involves a patient whose knee pain did not respond to TCM acupuncture but responded well to orthopedic acupuncture treating motor points, you could propose comparing different styles in observational studies to improve efficiency of care. If your patient with a recalcitrant condition responded well to a daily acupuncture regimen but not a weekly one, you might suggest reexamining treatment frequency as a parameter in research design to examine its impact on efficacy. Discussing specific ways to improve a research study may be much more fruitful than simply stating that more research ought to be done.

Recommendations pertaining to education and policy

It is unlikely that a single case report will change large-scale curriculum design in acupuncture colleges, but suggestions for educators may be supported depending on your case. For example, in a case focusing on the use of the 'surrounding the dragon' needle technique, a writer might suggest that this protocol be taught in all acupuncture colleges because it is so straightforward to apply.

As real changes in policy are dependent at this point on our current model of the evidence pyramid, it is unlikely that a case report can suggest significant changes to the scope of practice, laws, or insurance company reimbursement systems. However, it may still be useful to explore ideas as long as they are pertinent to the case at hand. With regards to discussion points that concern health care system inadequacies, one way to do so is to point out how the implementation of the ideas in your case report could improve any of the six qualities of good health care: safe, effective, patient-centered, timely, efficient, and equitable.[9] Touching on any one of these qualities may bring a broader perspective to the case and contribute to its overall significance.

Writing the Conclusion

A good conclusion should be centered around the main message of your case report. This final statement should be succinct and brief. Your task is to show how your work advances our understanding of acupuncture, as well as to bring the paper to a close by satisfying the original intent of the paper as expressed in the abstract. Many journals don't require a conclusion — it may be superfluous, since all of the significant points have been covered in the discussion. At this point you can at this point assume that your reader has read your paper fully and is looking for a focused and concise expression of the significance of your case. Please be sure to avoid making overgeneralized statements that are not supported by your one case and try to go beyond simply stating 'more research needs to be done'. Aim for 4–6 sentences where you summarize your discussion points and the lessons learned.

CHAPTER 11

Title, Abstract, References

Title

The title is one of the most important parts of your case report.[1] The goal of writing the title is to effectively and concisely communicate the article's contents to the reader. Due to the fact that most readers will find your case report through an electronic search engine or scientific database, it is best to keep the title straightforward and simple. To alert your reader to relevant content, the title should include the condition, the treatment modality, and the words 'case report,' as in the title 'Treatment of Fibromyalgia with Acupuncture: A Case Report.' A case title can be longer if the case involves multiple conditions or multiple treatment modalities (e.g., 'Case Report: Treatment of Traumatic Brain Injury and Peripheral Neuropathy with Electroacupuncture, Moxibustion, and Chinese Herbal Medicine.') The key to creating a good title is to incorporate all three elements as concisely as possible.

The first essential element of a case report title is the condition of the patient in your case. This can be a biomedically defined condition as long as it is a diagnosis made by a qualified practitioner. For example, if your patient's primary care physician has diagnosed him with atrial fibrillation, this condition can be included in your title. But if the patient had heart palpitations and during the initial intake said "I think I have AFib" but was not diagnosed as such by a biomedical physician, it would be misleading to include atrial fibrillation in your title because his palpitations could be due to any of a number of biomedical diagnoses (e.g., atrial flutter, atrial tachycardia).

If a patient does not have a specific biomedical diagnosis, the title may include the name of a symptom or symptoms.[xiii] An example of a symptom being used as part of a title would be 'Treatment of Insomnia with Moxibustion: A Case Report.' A group of symptoms may also be named, as in 'Case Report: Treatment of Infertility and Polycystic Ovarian syndrome with Acupuncture.' For case reports written primarily for an audience of acupuncturists, it is also acceptable to include traditional terminology as part of the title such as, 'The Treatment of Restless Organ Disorder (*Zang Zao*) with Acupuncture: A Case Report.'

The second key element of a case report title is to state the treatment modality or modalities; this may be as simple as 'acupuncture.' Depending on your audience, you may want to define your style of acupuncture more specifically. For example, if you intend to include your case in a volume that showcases a variety of acupuncture styles, you may want to specify which one you used. Depending on your case, you may need to include additional modalities such as moxibustion or laser acupuncture. Alternatively, the treatment modality may describe a more specific aspect of traditional theory: 'Using External Dragons in the Treatment of Generalized Anxiety Disorder: A Case Report.'

Unless your title captures the essence of your case, the people who might benefit from reading it may not be able to find it. If your case report is about a patient with rheumatoid arthritis but you only note arthritis in your title, it might be overlooked by someone doing a focused search. According to Lang, "the title may be considered the most important part of a scientific article. It is the most often read and often the only part read."[1] With such vast numbers of articles and information now

---

xiii One reason why a potential patient might seek out an alternative or complementary therapy such as acupuncture is that her doctors may not have been able to match her symptoms to a specific biomedical diagnosis.

available, being precise will allow potential readers to find your case and also save time for readers who are looking for something else.

There are a few caveats to keep in mind when writing a title. First and foremost, avoid clever titles relying on humor or mystery to draw the reader in. These titles generally do not inform the reader about the contents of the case, and few readers will be able to appreciate the humor or solve the mystery without reading it. For example, two titles that appeared in the *British Journal of Radiology* and the *Journal of Medical Imaging and Radiation Technology* were "An unusual incidental finding" and "*Primum non nocere*: first, do no harm" (respectively). Neither of the titles indicate that they are actually case reports.[2]

Also avoid writing an interrogative title, a title in the form of a question. Journal articles that end in a question mark are most often editorial articles or opinion pieces rather than case reports.[1] For example, 'Can Acupuncture Treat Diabetic Neuropathy?' might be a review of the research literature, or possibly a discussion of potential mechanisms, but not likely a case report. A third type of title to avoid is a declarative title or a title that contains a verb. For example, for 'Acupuncture Cures Headache In Just One Treatment' and 'Chinese Herbal Therapy Successfully Treats One Case of Parkinson's Disease,' each of these titles have a sensational type of sentiment because they each contain a verb. Many clinical journals do not allow declarative titles.[1] Finally, avoid abbreviations in titles; make sure that you spell out the full term. Normally, the first time it appears within the text of the article, it is also spelled out fully again, followed by the abbreviation in parentheses.

Abstract

The abstract is a concise summary of the case, where the goal is to attract readers and communicate the highlights. Aside from the title, the abstract is the next most important element that orients your reader to the contents of your case report. When searching a database for relevant items, a potential reader who is drawn in by your title would then likely read the abstract to determine whether or not to continue and read the article itself. Abstracts are typically 150–250 words long depending on the publication.[3] They include a summary of the introduction, the case presentation, treatment, and outcomes, concluding with a brief statement on the significance of the case.

Write the abstract after you have completed each of the other sections, as it is easier to write a summary of a complete work. Although short and concise, your abstract must be wholly understandable on its own. Abstracts generally come in one of two formats: structured abstracts, which are presented with a series of subheadings, and informative abstracts, which are written in paragraph form. Whether you choose to write a structured or informative abstract will depend on the conventions of the journal you submit to. If you write your case report as an academic assignment, your course instructor will note which type of abstract to use.

Structured Abstracts

Structured abstracts are found primarily in clinical journals where information is presented under a series of headings. Typical headings include: Background, Objectives, Patient, Methods, Outcomes, and Conclusion. Each journal typically has its conventions regarding the headings in the abstract. If you are submitting for publication in a journal, make sure that you read several case reports from the journal's past issues to learn the preferred abstract format, but keep in mind that even within one journal, you may find variation in the number and types of headings.

The Background subheading (sometimes labeled Introduction) is a one or two sentence summary of that section of your case report. It should not include any citations; if you need to cite a source, you are likely including too much detail. Any abbreviations must be spelled out in their entirety the first time they appear in the text, regardless of whether or not they are spelled out in the title. Depending on the journal, you may include a brief definition of the condition and relevant treatment, as in 'Amyotrophic Lateral Sclerosis (ALS) is a condition characterized by . . . Biomedical treatment for this condition is . . . Acupuncture has shown promise in clinical studies as a treatment for . . .' Depending on the journal, a sentence on relevant research may also be appropriate.

The Objectives subheading is more appropriate for clinical trials or basic mechanism research; however, since the format in many journals includes this subheading, a few examples are listed below:

i. 'To describe one case where intractable insomnia was treated successfully with acupuncture.'

ii. 'To examine the effect of acupuncture and moxibustion on a patient with chronic diarrhea.'

iii. 'To demonstrate the successful application of Five-Phase theory in the acupuncture treatment of a patient with panic disorder.'

The Patient subheading should include the patient's age, gender, condition(s), and symptom(s). It is reasonable to include a few words describing the severity of the condition, so that you may have a baseline to compare the results to. Be sure to avoid including any unnecessary information which may compromise the patient's identity. Here are a few examples:

i. 'The patient was a 22-year-old male who had constant severe pain radiating from the neck to the left arm as a result of a motor vehicle accident that occurred over three years before the onset of treatment.'

ii. 'The patient was a 45-year-old female with migraine headaches, including visual aura and nausea, occurring approximately twice per month, each lasting two days on average.'

iii. 'The patient was a 3-year-old male with weeping eczema that gradually became more severe during the four months immediately preceding treatment. He had four lesions on the limbs, each approximately 4 cm in diameter.'

The Methods subheading (sometimes labeled Interventions) should name the interventions used in the treatment of your patient, including the style of acupuncture along with the frequency and duration of the treatment course. There may be more than one intervention, as seen in the second example below:

i. 'The patient received eight weekly acupuncture treatments in the style of Kiiko Matsumoto, followed by two additional treatments over the following four weeks. In total, ten acupuncture treatments were administered over the course of twelve weeks.'

ii. 'The patient received 5 auricular acupuncture treatments given on a daily basis, along with decocted Chinese herbal formulas and nutritional recommendations.'

Under the Outcomes subheading (sometimes labeled Results), outcomes should be described as they relate to the baseline at the time of the first patient interview, as the examples below illustrate. Any follow-up data beyond the last treatment may also be included.

i. 'The patient reported gradual improvement in the severity of his low back pain, to the point where he was able to resume all normal activity without discomfort.'

ii. 'The patient was able to conceive after three months of treatment. She delivered a healthy baby without complications at 39 weeks.'

iii. 'The patient did not respond to the first several treatments given in the TCM style of acupuncture, but after electroacupuncture technique was added, he felt gradual improvement in the severity of his lower back pain. At the initial intake, he verbally reported a pain level of 6/10, which decreased to 1/10 by the end of his treatment course.'

If your case was a bit more complex and you treated the patient in two or more stages (possibly due to changing the diagnosis or treatment at various points during the treatment course), you may briefly explain your thinking.

Lastly, under the Conclusion subheading, use one or two sentences to state the main significant points of your case. These sentences should be a summary of the discussion and conclusion sections of your case report, and therefore should include recommendations that you may have for education, research, and practice. For example:

i. 'Orthopedic acupuncture may be an effective treatment option to consider for patients with lateral epicondylitis. Future clinical research may be designed to test whether addressing both anatomical injury and TCM pattern imbalances can lead to a quicker recovery.'

ii. 'In this case, a gentle style of non-insertive Japanese acupuncture appears to have dramatically reduced the patient's peripheral neuropathy pain. Patients with poor wound healing and neuropathy may wish to pursue Toyohari acupuncture, especially as insertive needling into areas with poor circulation has associated risks.'

Due to the fact that you have only one case to draw from, be sure to avoid making any definitive statements that indicate a cause and effect relationship between treatment and outcomes.[3]

Informative Abstracts

The informative abstract is a summary of a case report that includes all of the information seen in a structured abstract but is instead in paragraph form. While the structured abstract is a bit easier for new case report writers because its format ensures that all salient details will be included, if your case is not as straightforward you will have more flexibility if you use an informative abstract format like in the example below:

This single case outlines the acupuncture treatment of a 34-year-old female patient with chronic pain from psoriatic arthritis. She presented with multiple painful and inflamed joints, including in the hips, shoulders, lumbar spine, jaw, fingers, and toes. She experienced significant pain, which interrupted her sleep nightly and kept her from focusing her attention on a daily basis. Concurrently, she and her partner hoped to conceive and were concerned with pregnancy risks associated with her pain medications. After 21 acupuncture treatments over the course of 14 weeks, the patient experienced relief to the extent that she was able to completely eliminate the use of pain medication before conceiving with her partner. At various stages in this case, a variety of acupuncture approaches were applied, including methods from Traditional Chinese Medicine, Five-Element constitutional acupuncture, and Japanese styles of acupuncture. This case demonstrates the clinical choices made in the application of various acupuncture modalities along this patient's treatment course. Using case reports such as this to discern the strengths, weaknesses, and best applications of each style may be useful in guiding the evolution of acupuncture practice.[4]

As you can see, this abstract contains all of the relevant pieces contained in the structured abstract: the background (setting the context), the case description, methods, outcomes, and a short statement on the significance of the case.

Descriptive Abstracts

A third style of abstract, only occasionally seen in published journals, is the descriptive abstract.[1] This type of abstract is a very brief preview of the information contained in the article and does not contain any specifics about the patient, treatment, diagnosis, or outcomes, as seen in the following example:

> Irritable Bowel Syndrome (IBS) is one of the most common functional gastrointestinal disorders in North America. The severity and chronic nature of this condition have a significant impact on health-related quality of life. With few effective therapies available, there is a need for integrative approaches to symptom management. This report describes a successful case of using acupuncture and moxibustion to reduce symptoms of constipation-predominant IBS.[5]

While this type of abstract describes the case rationale, a reader will not gain much useful information from scanning it because important case details are missing. For this reason, readers generally prefer structured or informative abstracts rather than descriptive abstracts, as they make it easier to understand whether the article will be useful to them.

References

If you are submitting your case report to a professional journal, the required format for references will be contained in the *Instructions for Authors* (also called *Author Guidelines* or *Submission Guidelines*). If you are writing for a class, your instructor should indicate which format is required. Your reference list should include a variety of academic sources that can include theoretical articles, clinical trials, biomedical texts, traditional acupuncture texts and articles, research reviews, and more. Excessive referencing should be avoided; one publication provides a general guideline of aiming for no more than 15 references for a case report,[6] but this is not set in stone.

Some preliminary guidelines for selecting references are offered in chapter 8, which covers the Introduction section. The majority of your citations will likely be in the introduction, as its goal is to provide background from a variety of sources regarding what is currently known about the diagnosis and treatment of the condition from both biomedical and traditional points of view. In the case description, you may need to cite a reference if the diagnosis style is nonstandard TCM, and the treatment section will need references to support the actions and indications of specific points and techniques. The discussion section will also likely have references. Throughout your paper, you will need to indicate which statements are quoted and which statements have rephrased content derived from sources. In-line citations may be in the form of superscript numbers (e.g., AMA citation style) or in parentheses (e.g., APA citation style). If you are using AMA or any other citation format that requires superscript numbering for in-line text citations, you may find that keeping your citations in order of appearance is easy to lose track of, especially if you have done a lot of revisions. There are software tools that can help you to make sure your statements are attributed to the correct sources and formatted correctly.

CHAPTER 12

Publishing Your Case Report

THUS FAR, this book has provided a guide to writing each of the separate sections required in a comprehensive case report. While the order of sections as presented in this book provide a practical sequence for organizing one's thoughts during the writing process, the standard order of sections in a published case report paper is as follows: Title, Abstract, Introduction, Case Description (including Case History, Diagnosis, Treatment, and Outcomes), Discussion, Conclusion, and References. If you are writing a case report for an academic assignment, once you assemble these sections in the proper order and proofread them, your assignment should be complete.

In 2013, consensus-based clinical case reporting (or CARE) guidelines were published with the aim of disseminating and implementing systematic reporting guidelines for publishing case reports.[1] These guidelines, which differ slightly from the list above, were presented as a checklist of thirteen items recommended for inclusion in published case reports: 1) Title, 2) Keywords, 3) Abstract, 4) Introduction, 5) Patient Information, 6) Clinical Findings, 7) Timeline, 8) Diagnostic Assessment, 9) Therapeutic Intervention, 10) Follow-Up and Outcomes, 11) Discussion, 12) Patient Perspective, and 13) Patient Consent. Please note that the timeline is not typically required by publications but can be useful to clarify a complex case. Patient consent for de-identified patient information is not generally necessary for academic coursework but may be required for journal publication.[xiv]

---

xiv Additional guidance can be found in Appendix C where patient consent is discussed.

192

If you intend to submit your work for publication, first consider your case from the journal editor's point of view. A case report describing a patient with a straightforward condition who responded to a textbook treatment may not be of interest because it does not add significantly to the literature. If a case's reported outcomes are based solely on the patient's subjective experience without any specific measurement criteria, it may be rejected for poor validity. If a case's discussion points merely suggest that 'further research is necessary' without making specific suggestions towards changing research design for that condition, it may also be rejected. It may be helpful to discuss your case with a mentor or a colleague who has published, as he or she may aid you in adding significant discussion points or other improvements you overlooked, thus increasing the publication potential of the case.

Choosing a Journal

If your aim is to submit the paper to a professional journal, please familiarize yourself with the wide variety of journals that publish acupuncture case reports to determine which journal would be most appropriate.[xv] In general, they fall under two broad categories: biomedical journals and journals pertaining to Chinese medicine. The category you choose will depend on who you think will benefit most from reading your case report. If understanding your case requires a significant traditional theory background, or if your treatment involves more traditionally guided choices, you may want to submit your case report to a Chinese medicine journal whose readership has the more appropriate background for this content. Some Chinese medicine journals cater more to the TCM approach, while some journals are open to other styles.

If your discussion points delve into biomedical mechanisms or if your case includes a relatively straightforward protocol that might be effectively applied without adjustments to patients, you may choose a

---

xv Please consult my website (https://www.acupuncturecasereports.org/) for an updated list of journals that accept acupuncture case reports.

biomedical journal geared towards a biomedical audience. These may be biomedical journals with a specific subject focus, or biomedical journals dedicated to case reports. Biomedical journals solely dedicated to publishing case reports have increased in numbers in recent years; as of 2015, at least 160 peer-reviewed biomedical case report journals were produced by 78 publishers. At that time, 94% of these case report journals were open access, and 41% of the total number were indexed in the medical journal database PubMed, increasing exposure and potential to influence research and practice.[2] In recent years, some biomedical journals which historically focused on clinical trial research have established new companion journals that are more likely to contain case reports.

The business models of these journals vary; some journals charge a submission fee, a publication fee, or a membership fee to the author in order to provide free open access to all readers. Others charge readers a small fee per article accessed.[3] Please be aware: the open access publishing movement has given rise to an increasing number of 'predatory' journals that charge authors processing charges without providing peer review, editing, or preservation of contents. Some falsely claim to be indexed in PubMed or dishonestly claim to have a high journal impact factor. Be wary of journals that send email solicitations, promise acceptance decisions within a given time period that is too short for careful peer review, have an unprofessional appearance (e.g., grammatical errors, uneven typeset, dysfunctional web links, etc.), or lack a traceable editor-in-chief who has valid academic credentials.[2] If you do not recognize any individuals on the editorial board, do a quick web search for some of the names to see if they are legitimate (e.g., are members of academic faculty, or have published articles).

Revising for Publication

After choosing a journal whose readership may benefit most from your case report, thoroughly examine several sample case reports in that journal's recent issues to make sure that your case report adheres to its format. This is important because the overall presentation of case reports can vary somewhat from journal to journal. For example, one journal's articles may contain structured abstracts with subheadings, where others might prefer informative abstracts in paragraph form. The section following the abstract might be called the 'introduction' section in some journals, and in others it might be the 'background' section. The introduction section might be divided into two subheadings (as the biomedical introduction and the acupuncture introduction), or these two portions may be combined into one longer uninterrupted section. The journal may or may not expect a conclusion section. Not all journals require all sections — depending on the journal, significant editing on your part may be required to fit its conventions. Note that the case reports will likely be consistent in format, although you may find discrepancies between case reports even in the same journal.

If you intend to submit your case report to a specific journal, you can save time by tailoring your initial draft to match the conventions and audience of this target journal. For example, if you are submitting to a biomedical journal where recently published case reports omit the traditional acupuncture portion of the introduction section, you do not need to include that material in your draft. If the biomedical journal does include a small amount of traditional perspective in its articles, make sure you present any acupuncture concepts or terminology in a way that can be easily understood. If the diagnoses in a journal's published case reports are sparse on traditional explanation, you may presume that the readership may not have much background knowledge, thus leading you to simplify your case description, diagnosis explanations, and treatment rationales. Due to word count constraints and interest of the readers, translating and explaining each theoretical term and concept may not be necessary for the journal you have chosen.

If you are submitting your work to a biomedical journal, make sure that your outcome measures are sufficiently objective; to do so, it may be beneficial to collaborate with the patient's biomedical physician. Also, recognize that if published, your protocol may be followed without modification by allopathic or biomedical practitioners with no traditional background. If you are concerned about this, you may want to include a statement about acupuncture as an individualized and patient-centered treatment modality: 'The acupuncture treatment administered in this case may not be uniformly suitable for use in all patients with this condition. Modification based on clinical presentation may be required.'

In addition to revising for style and audience, your case report will also need to conform to the format and length of similar articles in the selected journal. Read and follow the *Instructions for Authors* carefully. It contains numerous specifics such as word counts, proper format, and reference style. The length of a published case report may vary greatly from journal to journal. If you are planning from the start to submit to a specific journal, check your word count periodically as you write to make sure that you are not getting bogged down by providing more background detail than is necessary. If you find that your paper exceeds the requested word count, one of the most flexible areas to condense is the introduction section; edit carefully while still making sure that all important aspects are included.

Finally, thoroughly proofread your paper. If you are submitting to a journal for the first time, it is recommended that you have a colleague, preferably one who has published in the past, help you proofread your work for errors or omissions. Do not rely solely on the grammar and spelling check functions on your word processor or apps like Grammarly, which may fail to catch errors. If you are not confident with English as a first language, consider asking a trusted colleague to help you proofread for usage and grammar.

Keywords, Authorship, and Citations

While the preceding topics covers the essential elements needed for publication, there are several additional features that you may want to consider.

Keywords

Some journals will request keywords with article submissions. Keywords are search terms that will be used to enable search engines to call up your case when a potential reader is looking for relevant material. Some journals will have a preset list of acceptable keywords; some keywords can be terms that consists of two or three words (e.g., rheumatoid arthritis, carpal tunnel syndrome). Because the primary medical search engines are biomedical, keywords are often drawn from the MeSH (Medical Subject Headings) list, issued by the National Library of Medicine.[xvi] The keywords will often include the name of the condition and the name of the treatment(s). If one of your keywords is an acronym, be careful to spell it out rather than using the acronym because search engines may omit acronyms as search terms. Please consult the *Instructions for Authors* (also known as *Author Guidelines* or *Submission Guidelines*) for the acceptable number of keywords and peruse several case reports in the target journal for examples. If you are limited to a specific number of keywords, choose the most relevant ones to your manuscript.

Authorship

If you are the sole provider managing the patient in the case report, you are likely the sole author. A case report may have more than one author if a second person contributed in a substantive way to the publication of the report. Any co-authors will need to agree to have their name in print as a co-author and will also need to approve the final version of the paper before it is submitted. According to the Guidelines of the International Committee of Journal Editors, the three criteria for being included as a co-author are:

1) has provided substantial contributions to conception and design, or acquisition of data, or analysis and interpretation of data;
2) has drafted the article or revised it critically for important intellectual content; and
3) has given final approval of the version to be published.[4]

For example, if you collaborated with a biomedical practitioner and discussed with them special considerations regarding the lab tests that

---

xvi See https://www.ncbi.nlm.nih.gov/mesh/ for a complete list.

were done, the biomedical practitioner could be considered a co-author. If another writer helped you to draft the article or revise it for content purposes, this also potentially qualifies the writer as a co-author.[5] In general, one would not expect to see more than a few authors on a case report.[xvii] If the three criteria above are not met by someone who contributed to the paper, they may be thanked in the acknowledgments at the end of the paper. Individuals to consider in the acknowledgments can include teachers who have taught you the specific and unusual techniques that your case is based on. Obtain consent from those you intend to acknowledge for their name to appear in print.

Citations

Check the *Instructions for Authors* to find the preferred citation style. If your case report is an academic assignment, your instructor should specify the citation style. The format of the reference list at the end of the paper should be checked carefully, and each statement within the text of the article which is cited should be indicated by an in-line text citation (most likely a superscript number).

Evaluating Case Reports

A rubric, as a systematically graded evaluation of academic work, can be an effective part of the teaching and learning process as well as a useful tool for assessing the quality of a case report. Specific and detailed feedback on the multiple elements clearly defined within a rubric may help a reader to identify the strengths and weaknesses of a complex case report. An instructor may use a rubric not only to assess student papers after

---

xvii Authors may need to sign a disclosure of conflict of interest and/or a financial disclosure form acknowledging all relationships and activities that might bias or be seen to bias their work. The conflicts of interest may be in the form of relationships with government agencies, foundations, commercial sponsors, and academic institutions, where the type of relationship could be financial (e.g., royalties, consulting fees, lectures, employment, honoraria) or nonfinancial (e.g. sponsored travel, administrative support, writing assistance, equipment).[6]

they have been written but also to explain expectations before students begin writing. A rubric can also be used by a case report writer in the form of self-evaluation.

Since 2005, the doctoral faculty at the Oregon College of Oriental Medicine (OCOM) in Portland have used a case report assessment rubric to evaluate the quality of student case reports.[xviii] While the CARE checklist and the OCOM Case Report Rubric as evaluation tools have similar aims of assisting authors in ensuring the completeness of a case report paper, there are some significant differences that make each assessment valuable to case report writers for different reasons. The CARE checklist can help a writer or evaluator ensure that each of the elements has been included, while the Case Report Rubric evaluates how complete the information is within each section and how effectively the writer communicates.

Peer Review

After you submit your paper, you should receive confirmation (usually by email) that your paper has been received. The editor-in-chief will read your case report to see if the subject matter and general quality fits with the journal's expectations. Most journals state in the *Instructions for Authors* that you may not submit your paper to any other journal while your paper is being considered for publication — doing so is generally considered unethical. Submitting your paper to only one journal at a time prevents the same paper from being published simultaneously in more than one journal, thereby avoiding a copyright violation. Submitting to one journal also respects the time and resources that the journal expends to evaluate and consider your work.

Many well-regarded journals use a process called peer review to determine the merits of a submitted paper. Peer review is a process of subjecting an author's work to the scrutiny of scholars and experts in the same field. Proponents of the peer review process maintain that peer review acts as a

---

xviii The full rubric is included in Appendix E.

filter to prevent low-quality or inaccurate manuscripts from reaching the scientific community. The goal is to make sure that published work maintains a high academic standard by avoiding publication of unwarranted claims, unacceptable interpretations, or personal views. Editors often aim to select peer reviewers who have published similar works in the past. Peer-reviewed articles provide a trusted form of scientific communication and reviewers donate their time and energy to raise the standard of the profession to ultimately help to improve patient care.[7]

That being said, in recent years the peer review process has been subject to criticism. Some of the criticisms of the peer review process are that it can be slow, expensive, time-consuming, subjective, inconsistent, and prone to bias. A 2002 systematic review of evidence concluded that peer review is "largely untested and its effects are uncertain."[8] Several suggestions for improving peer review include blinding reviewers to the identity of authors or, alternatively, eliminating blinding to increase accountability.[9]

Although the concept and the goals of peer review are fairly universal, the process of peer review may differ from journal to journal. One or more reviewers are chosen by the editor to comprehensively review your paper and provide comments regarding the validity, originality, quality, and significance of your work. Some peer review processes involve two types of comments. The first type is directed to the editor only, which contains the reviewer's opinion on whether or not to publish; these comments are confidential and are not revealed to the author. The second type of comments are directed to the author and are written to help the author know specifically what to revise and improve.

Ideally, a reader who is assigned to review the paper of an author he or she knows personally should recuse him or herself to avoid bias and a conflict of interest.[10] If you are asked to review an article for a journal, the journal should let you know if your identity will be revealed to the author. Whether or not your identity is revealed to the author, it is important for a peer reviewer to be collegial for the sake of the author; furthermore, the editor will also see these comments and will appreciate your professionalism.

Using the CARE guidelines as a checklist may certainly help to make sure that a submitted case report is complete. The Case Report Rubric can also be used as a more in-depth evaluation tool, which can be flexibly adapted to objectively evaluate not only the content but also its quality. It is worth noting that the discussion points are of primary significance in determining if the paper even makes it to peer review; if the teaching points are weak and include no significant practice, education, or research implications, it may be rejected out of hand even before reaching the review stage.

When the peer review process is complete, there are three possible outcomes. The first (which is quite unusual) is that the paper is accepted without any revision. The second is a flat out rejection without an opportunity for revision. The third possibility is acceptance with revisions required. If your piece requires revision, you will receive the comments of the peer reviewers and will then need to address these comments in your article. It is advised to provide a page of comments of your own to indicate how you addressed each reviewer comment. If you respectfully disagree with any of these comments, state your reasons and any further evidence you might have to support your point of view. Make sure you do not simply omit addressing the comment altogether or it will appear that you are being careless.[9] Once the article has been revised and sent back to the journal editor, the copy-edited version is sent back to the peer reviewers for final approval.

If you are successful in publishing your case report, the journal may contact you to see if you are willing to be a peer reviewer in the future. If your case report is rejected, do not be discouraged. A case report rejected by one journal may be warmly welcomed by another simply due to differing journal focus. If the reason for rejection is something that can be remedied, you may revise your case to address any inadequacies and submit it to a different journal. You can reduce the incidence of rejection by ensuring that your case report matches the journal's focus.

Final Thoughts

For many acupuncturists, writing case reports begins halfway through the internship year as an assignment to demonstrate the student's command of diagnostic skills, treatment planning, and patient care. As a result, case reports often have the feel of a stodgy academic exercise to be forgotten upon graduating into clinical practice.

More than 10 years ago, I began teaching a course on writing case reports for practicing acupuncturists in an advanced clinical doctorate program. My students ranged from recent licensees to practitioners with over 25 years of experience. Because many of them had learned from some of the profession's best instructors and mentors available before coming together in this program, they had a lot of experience to share with each other. Over the years, the cases which they wrote describing how they had successfully applied diverse approaches to a wide variety of conditions were inspiring. Many of these students have gone on to publish their work, and the status of the lowly case report has slowly begun to change from a mere academic assignment to a piece of evidence with the potential to change practice.

Not only can a case report offer novel solutions for a clinician puzzled about what path to take for a particular patient, but the ideas discussed can point the field in new directions. Cases go beyond simply generating research hypotheses to delving into new ideas in practice, whether it be understanding how to treat modern diseases in the context of ancient theory or revealing a technique or application not yet published in textbooks. Because case reports contain these clinical pearls within a practical context, the knowledge is presented in a way that encourages correct application. In clinical practice, part of finding an effective treatment is understanding the strengths of one approach while being aware of the limitations of others; following the logic in a case report helps the reader apply the right technique at the right time. The goal for writing case reports shifts beyond merely an exercise to prove oneself a capable student into a professional communication by a practitioner sharing insights and expertise with colleagues to grow the field of knowledge.

The growth of biomedicine is dependent on innovations generated by pharmaceutical and biotechnology companies, as the individual physician is generally discouraged from trying new approaches due to the high risk of adverse effects and ensuing legal consequences. Acupuncture does not advance in this top-down mechanical fashion. Because of acupuncture's favorable safety profile, individual practitioners are able to test new ideas on difficult cases and as such are the drivers of innovation. Master Tung, Kiiko Matsumoto, J.R. Worsley, and many others throughout the history of acupuncture have perfected their art case by case. We have been fortunate to receive the wisdom in their teachings and can add to this record by providing the evidence base that supports traditional practice, demonstrating how acupuncture can be performed successfully as we develop it further. While many of us might not yet have the insight or experience of the masters we revere, we can still learn from every case treated and share our insights with others. Once the case report literature has been fortified, through reviewing the aggregated evidence, our profession may be able to determine the most important techniques and approaches and work towards consolidating educational curricula and board exams to create stronger standards of practice.

By now, the reasons for writing case reports and the process of writing them should be clear. Through rigorously documenting our thought processes in patient care, we make a record of the current diversity of practice, while at the same time learning what we have in common. Through our writing, we can puzzle through whether and how to incorporate new technologies and biomedical knowledge in light of the traditional paradigm. And by describing how to apply theory in a way that generates effective treatment, we can both preserve traditional practice and move it forward into the future. It is our responsibility to write case reports to preserve the gifts we have been given, to develop the best of them through scholarly collaboration, and to create a record of these fascinating times for future generations of acupuncturists.

My hope is that this work will help you to share your experience with the broad community of practitioners and researchers to elevate the profession of acupuncture in our evidence-based world.

APPENDIX A

Acupuncture Styles

Chinese medicine is a rich and multifaceted discipline, with many currents of thought that weave throughout its history. Ideas from classical texts such as the *Huang Di Nei Jing* (Yellow Emperor's Classic), *Nan Jing* (Classic of Difficulties), *Shang Han Lun* (Treatise on Cold Damage) and the *Yi Jing* (Classic of Change) have been interpreted by many practitioners over time, giving rise to a variety of schools and methodologies in Chinese medicine. Historically, Chinese medicine has included acupuncture, Chinese herbology, *qigong*, martial arts, geomancy (*feng shui*), and Chinese astrology, all based upon traditional theories such as those describing qi, *yin/yang*, and *wuxing* (also known as five phases or five elements).

Acupuncture has diverged into multiple styles worldwide, characterized by a number of unique perspectives on diagnosis and treatment. This appendix includes descriptions of the major acupuncture styles practiced today.[xix]

TCM Acupuncture

As a result of political upheaval and pressure towards modernization in China, in the early 20th century many traditional theories were discarded

---

xix Special thanks to the many leaders in the field (see acknowledgments) who have helped write or consult on the accuracy of these descriptions.

and traditional practice was prohibited. Starting in the 1920s, a handful of traditional practitioners came together to revive traditional medicine, and together they formulated the system known as Traditional Chinese Medicine (TCM). TCM is now taught as a standardized curriculum of acupuncture and Chinese herbal medicine in the largest Chinese medicine universities in China. It is also the primary reference style in most acupuncture colleges in the U.S.

Traditional Chinese Medicine is a 20th-century methodology of diagnosis and treatment derived from traditional sources. It comprises a specific systematic approach to traditional acupuncture and Chinese herbal medicine. A TCM practitioner evaluates a patient using traditional skills such as asking, looking, smelling, listening, and palpation. The patient's constellation of signs and symptoms are then grouped into diagnostic patterns of imbalance which the practitioner uses to design an acupuncture treatment. Acupuncture points are chosen according to their theoretical actions (e.g., SP 10 *xuehai* nourishes Blood), indications for certain conditions (e.g., PC 6 *neiguan* treats nausea), classifications (e.g., *ying*-spring points clear Heat), and channel associations (e.g., GB 34 *Yanglingquan* removes obstruction from the Gall Bladder channel). Older adjunctive techniques such as cupping, *guasha*, and moxibustion may be incorporated into treatment, as well as modern innovations such as electroacupuncture.

A common complaint about TCM acupuncture is that the acupuncture channel model is not a strong factor in guiding treatment and that TCM treatment strategies overall seem more appropriate for Chinese herbal prescription. Indeed, as a result of the unique circumstances of its origin, many theories and techniques practiced in what are considered traditional styles of acupuncture may elude or exceed the boundaries of the TCM approach. That being said, the TCM methodology of pattern differentiation is often the first approach to acupuncture that students encounter, and while it is not necessarily the most effective approach to diagnosis and treatment in all cases, when seen as an amalgamation of various traditional theories, TCM may provide a solid background for exploration of other styles.[1]

Japanese Acupuncture

Many different thinkers and innovators have contributed to the various styles of acupuncture practiced in Japan. As a result, no homogeneous method of diagnosis or treatment prevails; however, the multiple approaches and styles of acupuncture practiced in Japan tend to share certain characteristics. Palpation often plays an important role in diagnosis; the palpation of points, channels, abdomen, and pulses are used not only to diagnostically detect imbalances, but these areas are checked again after treatment to verify outcomes. Treatment points are precisely located according to palpatory findings; an active point in a given individual may be located off the traditional measured location depending on where a practitioner might find abnormality in skin texture (e.g., a nearby point which feels more moist, cool, or more tense than the surrounding area). Needling tends to be lighter and shallower (although not always) and moxibustion is often used to stimulate acupoints.

Several styles reviewed below illustrate a variety of diagnostic and treatment practices of Japanese acupuncture.[2]

Meridian Therapy (keiraku chiryo)

Meridian Therapy, a neoclassical approach to acupuncture, was formulated in the 1930s by a group of acupuncturists who were unhappy with the biomedical influences on acupuncture that were politically favored at that time. They developed a practical and consistent treatment system based on principles from the *Nan Jing* (Classic of Difficulties). Meridian Therapy puts the focus of acupuncture on the meridians rather than on the individual points needled for specific symptoms. In Meridian Therapy, accurate diagnosis relies on the four examinations (looking, listening/smelling, asking, and palpation), with palpation considered the most important means of diagnosis. Special emphasis is placed on six-position pulse diagnosis to arrive at one of four primary patterns (*sho*). Some styles of Meridian Therapy begin with tonifying points associated with the five phases on the limbs to balance the *qi* in the meridians; others also use dispersion technique for meridian excess. After the meridians are balanced, symptomatic treatment is applied. Meridian Therapy thus addresses the underlying cause of a disease by resolving an excess or deficiency in the meridians before the symptomatic manifestations are realized.[3]

Toyohari Acupuncture

Toyohari acupuncture developed in Japan in the second half of the last century and was influenced primarily by blind acupuncturists. Diagnosis of a primary pattern (*sho*) is derived from the four examinations, case history, abdominal diagnosis, and pulse diagnosis; treatment points are guided by five-phase theory, similar to Meridian Therapy style. One key difference with other Japanese styles of acupuncture is that in Toyohari style the needles are often not inserted; rather, when tonifying, the qi is manipulated at the surface of the skin with the tip of the needle only very lightly touching or not touching the skin, using the non-needling hand as a guide. A silver needle (40mm, 0.18mm gauge) is the preferred tool for this technique. Supplemental branch treatments include 'naso' (treatment of the region of the supraclavicular fossa), 'muno' (treatment of the inguinal region), 'kikei' (a unique extraordinary vessel treatment method), and 'shigo' (a method that employs the theory of 'midday-midnight'). Adjunctive techniques also used include thread moxa (*okyu*) and microbleeding. In Toyohari style treatment the focus of is on bringing balance to the whole system within the primary pattern of the patient as the root treatment while using branch treatments to support the essential balancing and treat specific symptoms. Pulses and abdomen are checked after needling to verify the treatment benefit.[4]

Shakujyu Therapy

Shakujyu Therapy was founded by Shoji Kobayashi in the late 1970s. This style maintains that symptoms are a result of disruption of the qi flow in the body, with fundamental Cold (*hie*) causing the decline of the body's vital force (*jing qi*). *Shakujyu* is a phenomenon mentioned in two chapters in the *Nan Jing* (Classic of Difficulties), referring to tight or painful areas of stagnation palpable in the abdomen, where *Shaku* is in a deeper layer of the body and *Jyu* in a more superficial one. These accumulations are observed in one of five mapped zones on the abdomen that correspond to each of the five primary yin organs (Heart, Spleen, Kidney, Liver, Lung). The primary diagnostic emphasis in this style is on abdominal palpation to discover the location of stagnation; pulse diagnosis is also used to determine a pattern of imbalance. Contact needling is a technique used to reduce stagnation and supplement *jing qi* vacuity through the stimulation of numerous points within a region. The technique is applied with a teishin, a special style of silver non-insertive needle (SJ needle), on the abdominal and dorsal regions associated with the organs needing treatment. Other supplemental techniques used in this style include gentle insertive needling, intradermals, direct moxa (*okyu* rice grain size), warming moxa (direct cones on the skin removed at the maximum level of comfort), moxa on the handle of the needle (*kyutoshin*), and bleeding technique. During the acupuncture session, abdominal, dorsal, and pulse findings are frequently assessed to verify beneficial change.[5]

Kiiko Matsumoto Style Acupuncture (KMS)

Developed by Kiiko Matsumoto, a living master acupuncturist, her style combines ideas and techniques of her teachers (including Yoshio Manaka, Kiyoshi Nagano, and Yoshihiro Kawai) with her own clinical ideas and experience. The theoretical basis behind this acupuncture style is rooted in the classical Chinese texts, with interpretations of classical medical language that demonstrate early awareness of modern biomedical concepts. This style often seeks to address significant events in the patient's history which may have been important in the development of the current condition (e.g., trauma, surgery, significant past illness).

In KMS acupuncture, diagnosis is made primarily through palpation. Matsumoto has identified hundreds of acupuncture reflex zones, each indicating a specific structural or constitutional imbalance. Reflex zones related to the patient's history are tested by palpation, and a reflex zone is active if characterized by unusual texture or sensitivity with pressure. If pressure on a related distant point changes abnormal findings at an active reflex zone, the distant point can be used to correct the associated underlying imbalance. Many of Matsumoto's reflex zones and point combinations are derived from the traditional names of acupuncture points. Needling tends to be light and shallow, and treatment often features Japanese style moxibustion (*okyu*) with a protective barrier ointment (*shiunko*). Often, several reflexes are addressed simultaneously in treatment, depending on the history and presentation of the patient. Reduction of positive reflex areas corrects underlying root imbalances, thereby resolving the patient's symptoms.[6]

Shonishin (Japanese Pediatric Acupuncture)

Shonishin (literally translated as 'children's needle') is a Japanese style of pediatric acupuncture that dates back to the 17th century but has become more widespread in Japan in the late 20th century. Shonishin for babies and very small children does not use regular acupuncture needles, but rather a variety of tools that are tapped, rubbed, stroked, or pressed onto the body surface as a gentle noninvasive treatment approach. One interpretation of classical texts holds that the channels and acupuncture points in children are immature and of a different nature, so these tools are applied over areas of the body surface rather than targeted to specific acupuncture points. Using pulse or abdominal diagnosis is difficult for infants and small children, so diagnosing one of four primary vacuity patterns (*sho*) is often achieved by asking the parents about a symptom picture. The skin texture is assessed before, during, and after treatment as a way to verify improvement and gauge when the treatment is complete. For older children, dermal needles, moxibustion, cupping (non-fire, less pressure and short periods), and gentle bleeding techniques are used.[7]

Balance Method Acupuncture

Popularized by the late Richard Teh-Fu Tan, Balance Method is a style of acupuncture based on the classical theory described in the *I Ching* (Classic of Change, or *Yi Jing* in pinyin). In this text, the fundamental relationship between *yin* and *yang* gives rise to the eight trigrams and the *ba gua*, and these concepts guide the selection of effective acupuncture points. One hallmark of Balance Method acupuncture is the recognition of relationships between acupuncture channels; these relationships are derived from correspondences between pairs of *I Ching* trigrams on the *ba gua*. Each acupuncture channel has relationships with several other meridians. For example, the Large Intestine channel is related to the Lung channel by interior/exterior pair *(biao-li)*, related to the Stomach channel by its anatomical location (both are considered *yangming*), related to the Kidney channel by its opposite place on the daily meridian clock, and related to the Liver channel due to the relationship between *yangming* and *jueyin*. To treat a problem on the Large Intestine channel, effective points can therefore be found on the Lung channel, Stomach channel, Kidney, or Liver channel, and a Balance Method practitioner selects effective points on the related channel through holographic imaging (described in Chapter 2). A variety of holographic images exist, and given the variety of images and channel correspondences, a particular problem may have many possible solutions.[8]

Tung's Acupuncture

Tung's acupuncture is a family lineage style, passed down from father to eldest son for many generations in accordance with tradition. The last descendant of the Tung family to practice, Tung Ching Ch'ang (1916-1975), known as Master Tung, decided to train students outside his own family. Master Tung organized his acupuncture points by region and listed actions and indications in his 1973 text but offered no theoretical basis for their use. The vast majority of these points do not coincide with points of the twelve regular meridians. In 1980 one of Master Tung's direct disciples, Dr. Wei-Chieh Young, identified 393 Tung acupuncture points with 206 distinct point names and came up with theoretical reasoning explaining the functions of Tung's points; a number of Dr. Young's principles now characterize the practice of Tung's acupuncture. Some point functions are explained by the channel relationships between *taiyin* and *taiyang*, *shaoyin* and *shaoyang*, and *jueyin* and *yangming* based on the open, close, and pivot relationship in the *Huang Di Nei Jing* (Yellow Emperor's Classic of Medicine), while other channel relationships, including those related through *biao-li* (interior-exterior) and their position on the daily meridian clock, can also be used to explain Tung point functions. The *taiji* holography locating method describes the use of holographic images in varying sizes and locations to explain the function of a wide variety of Tung points.

Several needling techniques and principles enhance the effectiveness of Tung acupuncture points, with the direction and depth of the needle insertion being key to its effectiveness. The Body Tissue Correspondence principle maintains that better effects can be attained when the needle tip stimulates the type of tissue related to the complaint being addressed. Using the *Dong Qi* (activate the *qi*) needling technique, the patient moves the painful limb to verify effectiveness while distal needles are manipulated. The *Dao Ma* (coupling) needle technique involves needling two or three consecutive needles to increase the treatment effect. The *Qian Yin* (guiding) technique combines a treatment point with a distant guiding point to enhance and focus its effect. Bleeding technique applied to certain Tung points may be used to rapidly relieve complaints due to Blood stasis.[9]

Microsystem Acupuncture

Microsystem acupuncture is based on bioholographic theory, where the characteristics of the whole are represented in each of its smaller parts. A map of the whole body is superimposed onto the microsystem body part such that points on that map can inform the diagnosis and treatment of the areas they represent. The use of microsystems in acupuncture practice began in the 20th century, but the origins of microsystem acupuncture can be attributed to the central concept of macrocosm/microcosm, where the whole of a complex structure may be represented in miniature form. Numerous microsystems have been created by mapping the human body onto a variety of body regions including the ear, hand, feet, abdomen, back, neck, scalp, face, nose, tongue, wrist, ankle, arm, leg and eye. Many of the specific points within each microsystem are representative of corresponding anatomical structures, but in modern interpretations, some points are characterized by references to Chinese culture and philosophy.[10]

Scalp Acupuncture

Scalp acupuncture therapy involves needling of specific locations and regions on the scalp in order to address health concerns. It is a relatively new therapy based on the traditional acupuncture theory, holographic theory, and modern neuroanatomy. The original scalp acupuncture system, developed in 1972 by Dr. Shunfa Jiao, was based on modern knowledge of the areas and functions of the cerebral cortex. By stimulating the scalp on a specific area of a given holographic map, the effect of the acupuncture is targeted to the corresponding part of the body. Needles are inserted obliquely and combined (if appropriate) with mild electrical stimulation, bleeding therapy, or moxibustion. At least ten distinct scalp acupuncture systems exist where a different representative image of the human body is mapped onto the scalp for each system. Scalp acupuncture is often clinically used to address central nervous system disorders, such as cerebral disease, maldevelopment of the nervous system, brain damage, and stroke sequelae.[10]

Auricular Therapy

Auricular therapy (also auriculotherapy) is a diagnostic and treatment system based on normalizing the body's dysfunction through stimulation of points and regions on the external earlobe (auricle). In 1957, French physician Paul Nogier introduced the inverted fetus map of the auricle, where an image of the fetus could be superimposed on the auricle to determine the location of effective acupuncture points. After many years of academic exchange between China and Europe, auricular maps were standardized in the 1980s. In clinical practice, diagnosis of imbalance can be performed visually (by detecting color changes, shape changes, papules, vascularization changes, dry or flaky skin) or by palpation (lightly palpating with fingers or instrument to assess sensitivity). Electronic devices can also be used as diagnostic tools to detect active points as locations of decreased electrical resistance. Treatment can range from applying techniques of acupuncture, electroacupuncture, acupressure, application of seeds or magnets, ultrasound, laser stimulation, cauterization, moxibustion, and/or microbleeding to specific points on the earlobe to bring about a healing response.[10]

Koryo Hand Therapy

In 1971, Korean acupuncturist Tae Woo Yoo formulated Koryo Hand Therapy by mapping the complete 14 meridian system (12 regular meridians, plus Du and Ren channels) onto the human hand to develop a hand micro-meridian system. In the Koryo hand system map, the middle finger represents the body's head, neck and trunk, the thumb and little finger represent the legs, the index and ring fingers represent the arms, the center of the palm represents the abdomen, and the dorsum of the hand represents the back. Koryo Hand Therapy uses a specific style of abdominal diagnosis, as well as *renying cunkou* pulse diagnosis, comparing radial to carotid artery pulses as described in Chapter 9 of the *Ling Shu* (Spiritual Pivot). Points are selected according to diagnosis to rectify channel imbalances as well as extraordinary meridian patterns, and treatment consists of stimulating points on the hand with pressure, magnets, needles, or moxibustion. The Koryo system is one possible microsystem map of hand points; several alternate hand maps not associated with Koryo hand acupuncture have also been noted in the literature.[11]

Sa Am Acupuncture

The Sa Am acupuncture style originated in the Joseon dynasty in Korea (1644–1742 CE). Sa Am, a Buddhist ascetic monk, conceived a strategy for treatment that combines a grouping of four acupuncture points to address *yin/yang* and five-phase imbalances. His students transmitted two versions of his four-needle technique, one for correcting excess or deficiency of a meridian or organ and a second for correcting Heat or Cold imbalances of a meridian or organ. These two strategies form the basis for a treatment approach common to acupuncture practitioners in Korea.

No specific diagnostic method is inherent to the style. Sa Am practitioners rely to varying degrees upon pulse, signs and symptoms, body type and other methods to discover the nature and location of the imbalance. Four treatment points are chosen from both the unhealthy channel and from channels related to it by the five-phase generation and control cycles. The channel points are chosen from among the *shu* transport categories (*he*-sea, *jing*-river, *shu*-stream, *ying*-spring, and *jing*-well) according to their phase association (metal, water, wood, fire, and earth). Each of the four points are needled with supplementing or draining technique to resolve the primary imbalance. Using this basic approach, many individual practitioners have created their own variations.[12]

Worsley Five-Element Acupuncture

Worsley Five-Element Acupuncture is a lineage rooted in ancient Chinese wisdom based on observation of the natural order of all life. J.R. Worsley (1923–2003) is credited with bringing Five-Element Acupuncture to the West. Associated with each of the five Seasons (Spring, Summer, Late Summer, Autumn and Winter) is one of five Elements (Wood, Fire, Earth, Metal and Water). These Elements are observed both in the natural world (the macrocosm) and within each individual (the microcosm). Each of 12 Officials including their qualities and responsibilities is associated with one of the 12 meridians of *qi* energy. Balanced harmony among the five Elements, along with the proper functioning of the 12 Officials, engenders well-being and health; imbalance manifests as symptoms of disease.

Central to Worsley Five-Element Acupuncture is the concept and practice of diagnosing and treating the Causative Factor, the root cause of imbalance within each individual. Every person exhibits a Color, Odor, inappropriate Sound, and inappropriate Emotion, from which a practitioner identifies the Causative Factor. Additionally, the practitioner assesses how well the three levels within each individual (Body, Mind, Spirit) are interacting. The treatments include point selection based not only on their association with the Causative Factor, but also on the Spirit of the Point, which reflects the inner resources of the individual; the spirit of each point is embodied in the acupuncture point name and the associated Chinese characters.[13]

Orthopedic Style Acupuncture

The specialty of Orthopedic style acupuncture has become very popular over the last 20 years. In this approach to treating pain and injury, the anatomically significant tissues are assessed and identified—this could be a muscle strain or damage of other structures, trigger points that have specific referred pain patterns, or poor posture and faulty movement patterns. Treating injury and restoring proper anatomical function is the goal of treatment, and diagnosis is performed through palpation, observation of posture and movement, orthopedic tests, and manual muscle testing.

Treatment involves needling of anatomically significant points including trigger and motor points, tendon and tendon sheaths, tendinomuscular junctions, ligaments, joint capsules, and neurologically active points. The role of the traditional acupuncture channels and collaterals (*jing-luo*) and organs (*zang-fu*) are also considered, and additional treatment strategies may be applied based on traditional theory. Treatment techniques may also include electroacupuncture, moxibustion, *guasha*, manual therapy, and rehabilitative exercise. In this integration of biomedical anatomy and orthopedics with traditional acupuncture, the stagnation of *qi* and Blood in pain and injury is usually defined by the anatomical tissue involved and secondarily merged with the meridians and the organs.

Specialized orthopedic acupuncture techniques and Chinese herbal medicine are used in sports acupuncture to address athletic injuries, enhance athletic performance, and assist the active patient in recovery from exercise and training. Additionally, the sports acupuncturist pays attention to the psyche of the athlete, so the body-mind-spirit approach of Chinese medicine is applied to the athlete and the mental and emotional states that result from training and competing.[14,15]

It is perhaps appropriate here to discuss dry needling, a hotly contested technique sometimes used in the context of Orthopedic style acupuncture. The term 'dry needling' was coined by Janet Travell in 1960 and describes the procedure of inserting a needle without injecting any fluid. Often used for treatment of musculoskeletal pain, dry needling involves needling trigger points and motor points. Muscle motor points have been defined as locations where the motor nerves enter the muscle belly, and myofascial trigger points have been defined as tender hyperirritable locations within muscles. (It is worth noting that a fair number of trigger and motor point locations coincide with acupuncture point locations, and those which do not coincide could be considered *ashi* points.) Since dry needling is an acupuncture technique and is used among other techniques in the context of orthopedic style acupuncture, dry needling is not considered a 'style' of traditional acupuncture. Orthopedic style acupuncture uses the technique of dry needling but also works within a comprehensive system of medicine far more complex than needling taut muscle bands.

Because these points are associated with a specific physiological activity, and because motor points are based on anatomical structures rather than the traditional channel system, dry needling has particularly attracted the attention of biomedical practitioners. A number of continuing education companies offer minimal dry needling courses to non-acupuncturists, including chiropractors, massage therapists, registered nurses, physician assistants, and physical therapists. In several states, legal proceedings have begun to determine whether dry needling as a technique is within the scope of practice of some of these disciplines.[16]

Conclusion

This survey of acupuncture styles is not exhaustive and the diversity of approaches continues to evolve. Case reports describing the application of an individual style of acupuncture can be very useful to readers. As more case studies are written and compiled, trends may emerge in which conditions respond more quickly to particular styles, or common factors become identified in terms of patient presentation which might lead to the selection of one or more styles over another.

Examples of Case Report Categories

Category	Examples
Patients with an unusual condition or presentation	McCann H. Stiff person syndrome. *Journal of Chinese Medicine.* 2008 Feb;86:12–17.
	Coura LE, Paterno JC et al. Electroacupuncture alleviates headache and reduces cerebral blood flow velocity in aneurysmal subarachnoid hemorrhage. *Medical Acupuncture.* 2016;28(2):100–104.
	Yamamoto T, Schockert T, Boroojerdi B. Treatment of juvenile stroke using Yamamoto New Scalp Acupuncture — a case report. *Acupuncture in Medicine.* 2007;25(4):200–202.
	Gallo A. Neuropathic pain associated with Chronic Inflammatory Demyelinating Polyneuropathy treated with acupuncture. *Medical Acupuncture* 2014;26(5):295–297.
Patients with recalcitrant conditions	Sprague M and Chang JC. Integrative approach focusing on acupuncture in the treatment of Complex Regional Pain Syndrome. *Journal Alt Comp Med.* 2011;17(1):67–70.
	Inoue M et al. Pudendal nerve electroacupuncture for lumbar spinal canal stenosis — a case series. *Acupuncture in Medicine.* 2018 Dec;26(3):140–144.
	Lee H. Post-Traumatic stress disorder treated with traditional Chinese medicine: a case report. *Meridians Journal of Acupuncture and Oriental Medicine.* 2015;2(2)23–29.
	Bloch B, Vadas L, Reshef A, Schiff E, Kremer I, Haimov I. The acupuncture treatment of schizophrenia: a review with case studies. *Journal of Chinese Medicine.* 2010;93:57–63.

Category	Examples
Unique treatment approach	Davis R, Cochrane S. Case study: single point acupuncture using GB 20 to treat lower back pain characterized by leg length discrepancy. *Australian Journal of Acupuncture and Chinese Medicine.* 2016;10(2):38–42.
	Day J. Treatment of plantar fasciitis with the *yuan-luo* point pair: a clinical case report. *Journal of Acupuncture and Meridian Studies.* 2019;12(6):192–196.
	Legge D. Acupuncture treatment of chronic low back pain by using the *jingjin* (meridian sinews) model. *Journal of Acupuncture and Meridian Studies.* 2015 Oct;8(5):255–258.
	Kotlyar A, Brener R, Lis M. Use of Yamamoto new scalp acupuncture for treatment of chronic severe phantom leg pain. *Medical Acupuncture.* 2012;24(2):123–128.
	Lorenz C. Abdominal acupuncture for chronic low back pain: a case study. *Journal of Chinese Medicine.* 2018;116:38–42.
	Yurasek F, Martin B. The treatment of primary hypertension using plum blossom needle therapy. *Journal of Chinese Medicine.* 2013;101:37–42.
Challenging diagnosis	Sezgin Y. The acupuncture therapeutic approach in temporal arteritis vasculitis: a case report. *Journal of Acupuncture Meridian Studies.* 2018;11(3):116–118.
	Yu GJ. The treatment of constipation with Chinese herbal medicine: a case history. *Journal of Chinese Medicine.* 2017;115:42–46.
New medical technology	Chen Y, Guan J. Acupoint injections of autologous blood given to a patient who had psoriasis for 20 years made complete disappearance of psoriasis skin plaques in 6 months: A case report and underlining hypothesis. *Acupuncture and Electrotherapeutics Research International Journal.* 2017;42:113–119.
	Ton G, Lee LW, Chen YH, Tu CH, Lee YC. Effects of laser acupuncture in a patient with 12-year history of facial paralysis: a case report. *Complementary Therapies in Medicine.* 2019;43:306–310.
	Liang S, Christner D, Du Laux S, Laurent D. Significant neurological improvement in two patients with amyotrophic lateral sclerosis after 4 weeks of treatment with acupuncture injection point therapy using enercel. *Journal of Acupuncture and Meridian Studies.* 2011;4(4):257–261.

Category	Examples
Application of traditional theory	Blome K. Treatment of chronic tension-type headache with balance acupuncture: a case study. *Journal of Chinese Medicine.* 2017 Oct;115:25–29.
	Homan C. Acupuncture divergent channel treatment for cystic acne. *Meridians Journal of Acupuncture and Oriental Medicine.* 2016;3(4):31–37.
Integrative medicine	Flynn E. Electroacupuncture for pain relief and relaxation during oocyte retrieval: a case study. *Journal of Chinese Medicine.* 2010;93:33–34.
	Koprowski E. A case study: ghost point acupuncture a useful adjunct to psychotherapy, pharmacotherapy. *California Journal of Oriental Medicine.* 24(1):4–5.
	Hu WL, Chang CH, Hung YC, Shieh TY. Acupuncture anesthesia for complicated dental extractions in patients with lidocaine allergy. *The Journal of Alternative and Complementary Medicine.* 2009;15(11):1149–1152.
	Ganiyu SO, Gujba KF. Effects of acupuncture, core stability exercises, and treadmill walking exercises in treating a patient with postsurgical lumbar disc herniation: a clinical case report. *Journal of Acupuncture and Meridian Studies.* 2015;8(1):48–52.
	Haid M. Successful acupuncture treatment of xerostomia after chemotherapy and radiation therapy for head and neck carcinomas. *Medical Acupuncture.* 2015;27(3):235–238.
	Alban J. Acupuncture treatment of frequent urination and nocturia postprostatectomy. *Medical Acupuncture.* 2008;20(2)119–121.
Reporting adverse events	Kung YY, Chen FP, Hwang SJ, Hsieh HC, Lin YY. Convulsive syncope: an unusual complication of acupuncture treatment in older patients. *The Journal of Alternative and Complementary Medicine.* 2005;11(3):535–537.
	Rosted P, Wooley DR. Bell's palsy following acupuncture treatment — a case report. *Acupuncture in Medicine.* 2007;25(1–2):47–48.
	Lin CW, Wang J, Choy CS, Tung HH. Iatrogenic bullae following cupping therapy. *Journal of Alternative and Complementary Medicine.* 2009;15(11):1243–1245.
Unexpected effects (benefits) of treatment	Haysom-McDowell AJ, Loyeung B, Walsh S. Case study: *guasha* for shoulder pain with an unexpected outcome for restless leg syndrome. *Australian Journal of Acupuncture and Chinese Medicine.* 2017;11(1):26–30.
	Schulman D. The unexpected outcomes of acupuncture: case reports in support of refocusing research designs. *Journal of Alternative and Complementary Medicine.* 2004;10(5):785–789.

Category	Examples
Multiple treatment approach comparisons	Cheng I. Thawing the frozen shoulder — a case study and clinical recommendations for the use of acupuncture in treatment of adhesive capsulitis. *American Acupuncturist.* 2013;62:25–29.
	Chiu E. Psoriatic arthritis managed by multiple styles of acupuncture. *Meridians Journal of Acupuncture and Oriental Medicine.* 2016;3(4):17–22.
Case series	Dimitrova, A. Introducing a standardized acupuncture protocol for peripheral neuropathy: a case series. *Medical Acupuncture.* 2017;29(6):352–365.
	Shrikhande A, Schulman RA, Lerner B, Moroz A. Acupuncture for treatment of chronic low back pain caused by lumbar spinal stenosis: a case series. *Medical Acupuncture.* 2011;23(3):187–191.

Patient Consent

The International Committee of Medical Journal Editors (IMCJE) in Uniform Requirements for Manuscripts Submitted to Biomedical Journals state that "Patients have a right to privacy that should not be infringed without informed consent. Identifying information, including patient's names, initials, or hospital numbers, should not be published in written descriptions, photographs, and pedigrees unless the information is essential for scientific purposes and the patient (or parent or guardian) gives written informed consent for publication."[1]

There is considerable discrepancy in the requirements of various journals regarding the level of patient consent required to publish a case report.[2] In some professional journals, patient consent is currently still considered optional, depending on the circumstance. While some journals do not require patient consent for submission or publication, it would nonetheless be advisable for ethical purposes and for legal protection to obtain written consent. This is especially important if the case report includes highly sensitive information, such as a history of abuse, addiction, sexually transmitted disease, mental illness, or a rare condition. In the past, medical case reports were primarily available to a small number of subscribers to hardcopy medical journals, but today the sharing of professional resources via the internet has raised the risk of identification significantly.[3]

Outside of organized research studies, it is important to recognize that the patient has provided confidential information to the health care provider for the purpose of receiving health care,[xx] not for the advancement of medical knowledge. This is particularly notable for acupuncture case reports, as explaining a patient's traditional diagnosis may require significant levels of personal detail. Strongly consider excluding from case reports the following potentially identifying yet likely nonessential characteristics: date of birth, place of birth, place of residence (city/state), occupation, race, religion, and details about family and family history. As case reports should always be carefully de-identified prior to publication, direct identifiers such as name and record numbers of course would always be removed.

Two nuances to note include that if the patient is a minor, and consent is pursued for case report publication, parental consent is required; and, if the medical history of a patient refers to treatment received at a hospital, school clinic, or other organization, or the information and data to be published was collected pursuant to a research study, the institution's privacy officer or institutional review board should be consulted for approval.

Privacy Laws in each state and country differ with regards to patient privacy, therefore no one consent form is likely to offer full legal protection in all jurisdictions. However, a number of biomedical journals do for ethical purposes require the patient to sign a consent form as included in their *Instructions for Authors*. Some journals have required patient and practitioner consent forms, but others may request that you retain the patient's consent form and submit a signed practitioner form stating that you have obtained a signed consent from the patient and will maintain it in the event it is requested by the journal.

---

xx In the United States, Protected Health Information (PHI) is protected by the Health Insurance Portability and Accountability Act (HIPAA), including not only identifying information such as the patient's name, address and birth date, but also any information about the individual's physical or mental health, as well as treatment administered. (https://www.hhs.gov/hipaa/for-professionals/privacy/special-topics/de-identification/index.html#protected)

The Committee on Publication Ethics (COPE) document on best practices for ensuring consent for publishing medical case reports[4] offers ethical guidance for creating a patient consent form. For weblinks to examples of patient consent forms used by biomedical journals for publication of case reports, as well as to the Committee on Publication Ethics document on best practices, please visit www.acupuncturecasereports.org.

APPENDIX D

Chinese Herbal Medicine

While this book primarily concerns case reports pertaining to acupuncture, many acupuncture colleges include training on the prescription of Chinese herbal medicine. Both therapies are often used concurrently and there may be overlapping theoretical reasoning behind the treatment choices made in both areas. If your case report involves Chinese herbal treatment, there are two sections that you may need to expand: the introduction section and the treatment section.

If the herbal component comprises a lead role in the treatment, adding relevant information to the introduction section will be useful, especially as it relates to the biomedical research on the herbs prescribed. This information may come in the form of clinical trials where the herbal formula (or perhaps even a single herb within the formula) has been tested for efficacy for the condition being discussed. Mechanistic studies may also be cited if the formula (or herb) has a known physiological effect that may have bearing on the condition.

Within the treatment section, details about the formula and the herbs must be included so the reader will be able to reproduce the therapy if desired or modify it if necessary. If your treatment involves either prepared medicines or individualized formulas, include the name of the formula in Chinese pinyin and English in parentheses along with the treatment principle. 'The patient was prescribed *Suan Zao Ren Tang* (*Zizyphus* decoction) to address his pattern of Liver blood deficiency.' If there are any further considerations as to why this formula was chosen, they may be described here. Describe clearly how long the patient took the herbs and how compliant they were with taking them.

For prepared medicines, include the manufacturer as well as daily dosage (e.g., '8 tea pills three times per day, manufactured by Plum Flower'). If the herbs were administered topically, make sure this is indicated. Do not use medical jargon such as BID or PRN. Depending on how much you wish to discuss specific ingredients, you may or may not choose to list all ingredients with dosages or percentages.

For individualized formulas, after describing the rationale behind why the formula was chosen, describe the manner of administration, whether it be granule extracts or bulk herbs in decoction. Sources of the herbs should be noted, including the manufacturer if the herbs were granule extract (e.g., KPC) or tincture (e.g., Blue Poppy), or the seller if bulk herbs were used (e.g., Mayway). List all herbs (with pinyin and Latin in italics) in a table, along with individual dosages. If significant modifications were given, an additional column in the table may be added to indicate the action of each herb within the formula.

Then, in the text after the chart, state the instructions for administration. For example, 'Bulk herbs were decocted for 30 minutes, and the liquid was strained and divided into four portions; the patient consumed two portions per day with food' or 'Granule extracts were taken 4 grams three times per day mixed in warm water.' If you prepared a topical administration or an herbal soak, describe the preparation. You will also need to describe and justify any modifications made to adapt the formula to your specific patient (additions, subtractions, or significant deviations from standard dosages of individual herbs within the formula).

Case Report Rubric

The Case Report Rubric[1] has twelve separate elements, each of which describes essential content while also providing a detailed description of what constitutes acceptable or unacceptable levels of performance. The twelve elements are: 1) overall appearance and writing quality, 2) title and language 3) abstract, 4) biomedical introduction, 5) acupuncture introduction, 6) case history, 7) diagnosis, 8) treatment, 9) outcomes, 10) discussion, 11) conclusion, and 12) references.

Each element of a rubric addresses how well the writer has fulfilled the requirements of its given section. For each of the elements, the evaluator chooses one of four levels of performance: 1) unacceptable, 2) improvement required, 3) professional level, and 4) exemplary. Specific descriptions of what is required to achieve each of the levels are included in the rubric. Additionally, specific comments can be added for each of the twelve elements to indicate what is missing and needs to be added or improved, or what has been done well. By reading these comments, the writer should clearly understand how well they achieved the goals of writing a case report, while at the same time recognizing what changes are needed to improve any weak areas.

Case Report Rubric[xxi]

Element 1 — OVERALL APPEARANCE AND WRITING QUALITY

The document is professionally presented, as indicated by well-organized sections with clearly labeled headings and proper formatting (page numbers, margins, and use of standard fonts). The text is well written, as indicated by consistent tone and absence of significant spelling and grammatical mistakes. The text flows well and concepts are presented in a logical manner.

1 – Occasional spelling or grammatical errors, poor formatting
Lack of consistent professional tone, poor organization

2 – Minor spelling or grammatical errors
Minor inconsistencies in tone and/or poor formatting
Minor issues with organization and flow

3 – Well-organized with proper formatting; consistent tone
Very minor spelling or grammatical errors

4 – Well-organized with clearly labeled headings, tables if appropriate
No spelling or grammatical errors; consistent tone and proper formatting

Element 2 — TITLE AND LANGUAGE

Formulate an informative title, which orients the reader appropriately to the document's contents. Relevant medical vocabulary and terminology is used throughout the case report paper.

1 – Inappropriate or insufficient medical terminology used, unclear title

2 – Insufficient use of medical terms or unclear title

---

xxi The original version of this rubric was published in the *Meridians Journal of Acupuncture and Oriental Medicine*, now titled the *Journal of the American Society of Acupuncture (JASA)*.

3 – Relevant and sufficient use of medical terminology, appropriate title

4 – Publishable or nearly publishable quality, outstanding title

Element 3 — ABSTRACT

Summarize the overall contents of the case report by writing a concise abstract, establishing the rationale and/or purpose of the case report. It should include appropriate information on the Background, Objectives, Case Description, Treatment, Outcomes, and Conclusion.

1 – Substantial relevant information missing; or, provides different information or content than paper does

2 – All relevant information included with excessive extraneous material; or, some relevant information lacking

3 – All relevant information included with minor extraneous material, or minor relevant material missing

4 – All relevant information included succinctly with no extraneous material

Element 4 — INTRODUCTION/BACKGROUND—BIOMEDICAL

Write an introductory section establishing a biomedical context through an appropriate review of biomedical journal articles, texts and other research information. The biomedical background section should include biomedical information on the condition being discussed in the case, including diagnosis, demographics, etiology and pathogenesis, and treatment options.

1 – Superficial treatment of biomedical condition, possibly without treatment options

2 – Inadequate depth of coverage of biomedical condition; or, no journal articles referenced

3 – Missing minor aspects of condition or treatment, appropriate source material referenced

4 – Thorough description of biomedical condition with treatment options, good use of biomedical literature

Element 5 — INTRODUCTION/BACKGROUND—ACUPUNCTURE

Write an introduction section establishing a context through an appropriate review of acupuncture journal articles, texts and other research information. The acupuncture introduction should review the traditional perspective on the condition discussed in the case, including differential diagnosis, etiology and pathogenesis, and treatment options. Biomedical research should be reviewed for the condition being discussed, and articles from acupuncture journals with theoretical information on the subject should be included where appropriate.

1 – Superficial description of acupuncture perspective, possibly without treatment options; or, section reiterates textbook description of disease and treatment

2 – Inadequate depth of information for acupuncture perspective; or, inadequate source material

3 – Missing minor aspects of condition or treatment, adequate source material

4 – Thorough description of traditional diagnosis and treatment options with material cited from a variety of sources; thorough use of available literature

Element 6 — CASE DESCRIPTION—CASE HISTORY

Write a thorough narrative presentation of the patient's case history, including full details of the chief complaint, relevant past and present biomedical history, and traditional diagnostic information.

1 – Incomplete details on chief complaint; lack of traditional acupuncture diagnostic information, lacking in both traditional acupuncture and biomedical aspects of medical history

2 – Basic details on chief complaint; lack of adequate traditional acupuncture diagnostic information, OR lack of significant biomedical history information

3 – Detailed description of case history, including traditional acupuncture diagnostic information and biomedical history information

4 – Detailed description of case history, including traditional acupuncture diagnostic information, possibly biomedical lab values and their relevance to the biomedical diagnosis

Element 7 — CASE DESCRIPTION—DIAGNOSTIC ASSESSMENT

Provide a full assessment of the patient's traditional diagnosis, including disease categories, pathogenesis and etiology, patterns, and differentiations, as appropriate. The justification for your diagnosis and pattern differentiation is required (e.g., symptoms, signs, pulse, tongue, other information that supports your diagnosis).

1 – Traditional acupuncture diagnosis stated with no rationale

2 – Traditional acupuncture diagnosis stated with rationale but no discussion of pathogenesis, and etiology

3 – Traditional acupuncture diagnosis stated with rationale, pathogenesis, and etiology, but with minor omissions or inconsistencies

4 – Traditional acupuncture diagnosis stated with rationale, pathogenesis, and etiology

Element 8 — DESCRIPTION—TREATMENT

State your treatment principles and describe your treatment protocol as they relate to your diagnosis and pattern differentiation. Include the details on the point combinations and/or specific needling techniques used, and your justification for their use based on your diagnosis. Also include information on the case management—treatment duration and frequency, as well as adjunctive therapies (e.g., herbal medicine, diet, exercise, etc). Include a discussion of long-term case management and treatment strategies when appropriate. For cases with multiple treatments, treatments may be appropriately summarized.

1 – Acupuncture treatment stated without principles, and with no details; or, treatment principles and diagnosis from previous section do not match

2 – Acupuncture treatment stated with treatment principles, missing technical aspects of treatment

3 – Acupuncture treatment stated in sufficient detail so as to be performable by reader, including sufficient technical aspects of treatment and insufficient rationale behind treatment choices

4 – Acupuncture treatment stated in sufficient detail so as to be performable by reader, including sufficient technical aspects of treatment and rationale behind treatment choices. Good description of case management; adjunctive treatments described with adequate detail. Treatments are summarized appropriately if numerous.

Element 9 — CASE DESCRIPTION—OUTCOMES AND PROGNOSIS

Describe the outcomes of treatment, including signs, symptoms and tests that indicate progress (or lack of it). Also include the patient's prognosis.

1 – Outcomes briefly stated, with no objective markers, and not related to initial endpoint

2 – Outcomes with some detail, lacking prognosis

3 – Outcomes with adequate detail, prognosis is asserted with no basis

4 – Outcomes with adequate detail, prognosis is supported

Element 10 — DISCUSSION

Discuss your observations and the results in the case, summarizing the practical and theoretical points. Discuss how the case relates to the purpose you described in the introduction. Propose recommendations for clinical practice, case management, and/or further research based on this case.

1 – Results superficially reviewed and no significant analysis is presented

2 – Results are analyzed, and conclusions drawn

3 – Results are thoroughly analyzed, and conclusions drawn. One reflection on or recommendation for clinical practice is suggested

4 – Results are thoroughly analyzed, and conclusions drawn. Discussion brings up more than one recommendation for improving education, research, or clinical practice.

Element 11 — CONCLUSION

Summarize the practical and theoretical points of the case in a concise conclusion.

1 – Case superficially summarized or incomplete conclusions drawn

2 – Case is summarized, conclusions are drawn

3 – Case is summarized, conclusions are drawn, and practical or theoretical points are also summarized

4 – Case is thoroughly summarized, conclusions are drawn, practical or theoretical points are also summarized, and reference is effectively made back to the objective(s) and/or rationale for the case

Element 12 — REFERENCES

A good variety and sufficient number of professional sources are cited (not just basic texts). Citations are in proper format according to journal author instructions.

1 – Inadequate sources with improper format for citations; statements made in the case report which should be but are not cited; sources of poor quality

2 – Inadequate sources with minor problems with citation format; or sources of poor quality

3 – Section complete, with proper format of citations, both biomedical and traditional sources, but lacking in primary sources

4 – Section complete, with proper format of citations, both biomedical and traditional sources, with sufficient primary sources

References

CHAPTER 1

1. Thomas G. *How to do your case study*. Thousand Oaks, CA: Sage Publishing; 2016.

2. Nissen T, Wynn R. The recent history of the clinical case report: a narrative review. *JRSM Short Rep*. 2012 Dec;3(12):87.

3. Packer CD, Berger GN, Mookherjee S. *Writing case reports: A practical guide from conceptions through publication*. Switzerland: Springer International Publishing; 2017.

4. Furth C, Zeitlin JT, Hsiung PC, eds. *Thinking with cases: specialist knowledge in Chinese cultural history*. Honolulu: University of Hawaii Press; 2007.

5. Nyssen T, Wynn R. The history of the case report: a selective review. *Journal of the Royal Society of Medicine Open*; 2014 April:5(4) https://journals.sagepub.com/doi/full/10.1177/2054270414523410

6. Pomata G. Sharing cases: the *Observationes* in early modern medicine. *Early Sci Med* 2010;15(3):193–236.

7. Day R. The origins of the scientific paper: the IMRAD format. *American Medical Writers Association*. 1989;4(2):16–25.

8. Greene BN, Johnson CD. How to write a case report for publication. *Journal of Chiropractic Medicine*. 2006;5(2):72–82.

9. Pincus HA, Henderson B, Blackwood D, Dial T. Trends in research in two general psychiatric journals in 1960–1990: research on research. *Am J Psychiatry*. 1993;150:135–42.

10. Wilkinson G. Fare thee well — the editor's last words. *Br J Psychiatry*. 2003;182:465–6.

11. Agha R, Rosin RD. Time for a new approach to case reports. *Int J Surg Case RE*. 2010; 1(1):1–3.

12. Kidd MR, Saltman DC. Case reports at the vanguard of 21st century medicine. *Journal of Medical Case Reports*. 2012;6(article 156).

13. Cohen A, Stavryi P, Hersh W. A categorization and analysis of the criticisms of Evidence-Based Medicine. *Int J Med Inform*. 2004;73(1):35–43.

14. Cullen C. Yi'An (case statements): The origins of a genre of Chinese medical literature. In Hsu E, ed. *Innovation in Chinese medicine*. Cambridge, England: Cambridge University Press; 2001:297–393.

15. Furth C. Producing medical knowledge through cases: history, evidence, and action. In Furth C, ed. *Thinking with cases: specialist knowledge in Chinese cultural history*. Honolulu: University of Hawaii Press; 2007:125–151.

16. Nissen T, Wynn, R. The clinical case report: a review of its merits and limitations. *BMC Research Notes*. 2014;7(article 264).

17. Guyatt G, et al. Evidence-Based Medicine: a new approach to teaching the practice of medicine. *JAMA*. 1992 November;268(17):2420–2427.

18. Jenicek M. *Clinical case reporting in Evidence-Based Medicine*. 2nd ed. New York: Arnold Publishing; 2001.

19. Straus S, et al. *Evidence Based Medicine: how to practice and teach EBM*. New York: Churchill Livingstone; 2005.

20. Lukoff D, et al. The case study as a scientific method for researching alternative therapies. *Alternative Therapies*. 1998 Feb;4(2):44–52.

21. Britton A, McKee M, Black N, McPherson K, Sanderson C, Bain C.. Threats to applicability of randomized trials: exclusions and selective participation. *Journal of Health Services Research and Policy*. 1999 April;4(2):112–121.

22. Singal AT, Higgins PR, Waljee AK. A primer on effectiveness and efficacy trials. *Clin Transl Gastroenterol.* 2014 Jan;5(1):e45.

23. Gibbert M, Ruigrok WM, Wicki B. What passes as a rigorous case study? *Strategic Management Journal.* 2008;29(13):1465–1474.

24. Chronbach L, Meehl PE. Construct Validity in Psychological Tests. *Psychological Bulletin.* 1955;52:281–302.

25. Yin RK. *Case study research and applications: design and methods.* 6th ed. New York: Sage Publications Inc.; 2017.

26. Wifstad A. External and internal evidence in clinical judgment: the Evidence-Based Medicine attitude. *Philosophy, Psychiatry, & Psychology.* 2008 June;15(2):135–139.

27. Greenhalgh T. Integrating qualitative research into evidence based practice. *Endocrinol Metab Clin North Am.* 2002 Se;31(3):583–601.

28. Stux G, Hammerschlag R. *Clinical acupuncture: scientific basis.* New York: Springer Verlag; 2001.

29. Hsu E, ed. *Innovation in Chinese medicine.* Cambridge, England: Cambridge University Press; 2001.

30. Haynes RB, Devereaux PJ, Guyatt GH. Physicians and patients' choices in evidence-based practice. *BMJ.* 2002 Jun 9;325(7350):1350.

31. Reid T. The limitations and misuses of Evidence-Based Medicine: a critical evaluation. *Journal of Chinese Medicine.* 2015 June;108:15–31.

CHAPTER 2

1. Eckman P. *In the footsteps of the yellow emperor: tracing the history of traditional acupuncture.* San Francisco: Cypress Book Company; 1996.

2. Cassidy CM. Cultural context of complementary and alternative medicine systems. In Micozzi MS, ed. *Fundamentals of complementary and alternative medicine.* New York: Churchill Livingstone; 1996:9–34.

3. Wu NL, Wu AQ. *Yellow Emperor's Canon Internal Medicine.* Beijing: China Science & Technology Press; 2005.

4. So, J. *The Book of acupuncture points.* Brookline, MA: Paradigm Publications; 1989.

5. Birch S, Ida J. *Japanese acupuncture*. Brookline, MA: Paradigm Publications; 1998.

6. Baxter GD, Bleakley C McDonough S. Clinical effectiveness of laser acupuncture: a systematic review. *Journal of Acu and Merid Stud*. 2008 Dec;1(2):65–82.

7. Wakefield ME, Angelo M. *Vibrational acupuncture*. Philadelphia: Singing Dragon publishers; 2020.

8. Aldrich E, Bornemann R. *Fang xiang liao fa: essential oil analogues of TCM herbal formulas*. Self-published; 2013.

9. Deadman P, Al-Khafaji M, Baker K. *A manual of acupuncture*. East Sussex, England: Journal of Chinese Medicine Publications; 1998.

10. Shudo D. *Finding effective acupuncture points*. Seattle: Eastland Press; 2003.

11. Matsumoto K, Birch S. *Hara diagnosis: reflections on the sea*. Brookline, MA: Paradigm Publications; 1990.

12. Filshie J, White A. *Medical acupuncture*. New York: Churchill Livingstone; 1998.

13. Travell J, Simons D. *Myofascial pain and dysfunction: the trigger point manual: the upper extremities*. Philadelphia: Williams & Wilkins; 1983.

14. Matsumoto K. *Kiiko Matsumoto's clinical strategies*. Natick, MA: Kiiko Matsumoto International; 2002.

15. Callison M. *Sports medicine acupuncture*. San Diego: AcuSport Education; 2019.

16. Twicken D. *I ching acupuncture: the balance method: clinical applications of the ba gua and i ching*. Philadelphia: Jessica Kingsley Pub; 2012.

17. Head-Nugent A. Ashi points in clinical practice. *Journal of Chinese Medicine*. 2013 Feb;101:5–12.

18. Cheng X. *Chinese acupuncture and moxibustion*. 4th ed. Beijing: Foreign Languages Press; 2019.

19. Wang Y. *Micro-acupuncture in practice*. New York: Churchill Livingstone; 2009.

20. McCann H, Ross HG. *Practical atlas of Tung's acupuncture*. Germany: Muller & Steinicke; 2018.

21. Young WC. *Tung's acupuncture*. Taipei, Taiwan: Chih-Yuan Book Store; 2005.

22. Hicks A, Hicks J, Mole P. *Five element constitutional acupuncture*. New York: Churchill Livingstone; 2004.

23. Park M, Kim S. A modern clinical approach of the traditional Korean Saam acupuncture. *Evidence Based Complementary and Alternative Medicine*; 2015. https://doi.org/10.1155/2015/703439

24. Tan RT, Bensinger JW. *Dr. Tan's strategy of twelve magical points*. San Diego: R. Tan; 2003.

CHAPTER 3

1. Beck AH. The Flexner report and the standardization of American medical education. *JAMA*. 2004;291(17):2319–2140.

2. Duffy TP. The Flexner report — 100 years later. *Yale J Biol Med*. 2011 Sep;94(3):269–276.

3. Snyderman R., Weil AT. Integrative medicine: bringing medicine back to its roots. *Arch Intern Med*. 2002 Feb;162(4):395–397.

4. Reston J. Now, about my operation in Peking. July 26, 1971, *New York Times*, 1. https://www.nytimes.com/1971/07/26/archives/now-about-my-operation-in-peking-now-let-me-tell-you-about-my.html.

5. Pomeranz B, Chiu D. Naloxone blockade of acupuncture analgesia: Endorphin implicated. *Life Sciences*. 1976:19(11);1757–1762.

6. Liu YK, Varela M, Oswald W. The correspondence between some motor points and acupuncture loci. *American Journal of Chinese medicine*. 1975(3);4:347–358.

7. Travell J, Simons D. *Myofascial pain and dysfunction: the trigger point manual: the upper extremities*. Philadelphia: Williams & Wilkins; 1983.

8. Hui KKS, et al. Acupuncture modulates the limbic system and sub-cortical gray structures of the human brain: evidence from fMRI studies in normal subjects. *Human Brain Mapping*. 2000;9:13–25.

9. Hughes J, Smith TW, Kosterlitz HW, Fothergill LA, Morgan BA, Morris HR. Identification of two related pentapeptides

from the brain with potent opiate agonist activity. *Nature.* 1975 Dec;258:577–579.

10. Li G, Cheung R, Ma Q, Yang ES. Visual cortical activations on fMRI upon stimulation of the vision-implicated acupoints. *Neuro-report.* 2003 April 15;14(5):669–673.

11. Hong Y, Yoo Y, Han J, Wager TD, Woo C. False-positive neuro-imaging: undisclosed flexibility in testing spatial hypotheses allows presenting anything as a replicated finding. *Neuroimage.* 2019 July 15;195:384–395.

12. Elliott ML, et al. What is the test-retest reliability of common task-functional MRI measures? New empirical evidence and a meta-analysis. *Psychological science.* 2020;31(7):792-806. doi:10.1177/0956797620916786.

13. Rosenblum A, Marsch LA, Joseph H, Portenoy RK. Opioids and the treatment of chronic pain: controversies, current status, and future directions. *Exp Clin Psychopharmacol.* 2008;16(5):405-416. doi:10.1037/a0013628.

14. Vickers AJ. Placebo controls in randomized trials of acupuncture. *Evaluation & the health professions.* 2002 Dec;25(4):421–435.

15. Lao L, Ezzo J, Berman BM, Hammerschlag R. Assessing clinical efficacy of acupuncture: considerations for designing future acupuncture trials. In Stux G Hammerschlag G, *Clinical acupuncture: scientific basis.* New York: Springer-Verlag; 2001:187–209.

16. Long Y, Chen R, Guo Q, Luo S, Huang J, Du L. Do acupuncture trials have lower risk of bias over the last five decades? A methodological study of 4715 randomized controlled trials. *PLoS ONE.* 15(6):e0234491.

17. Ioannidis JPA. Why most published research findings are false. *PLoS medicine.* 2005 Aug. https://doi.org/10.1371/journal.pmed.0020124.

18. Prasad V, et al. A decade of reversal: an analysis of 146 contradicted medical practices. *Mayo Clinic Proceedings.* 2013;88(8):790–798.

19. Tatsioni A, Gonitsis NG, Ioannidis J. Persistence of contradicted claims in the literature. *JAMA.* 2007;298(21):2517–2526.

20. Prasad V, et al. A decade of reversal: an analysis of 146 contradicted medical practices. *Mayo Clinic Proceedings.* 2013;88(8):790–798.

21. Gluse NB. Riding the waves of change together: are we all paying attention? *J Med Libr Assoc.* 1997 Apr; 96(2):85–87.

22. Patsopoulos NA. A pragmatic view on pragmatic trials. *Dialogues Clin Neurosci.* 2001 Jun;13(2):217–224.

23. Singal AG, Higgins PD, Waljee AK. A primer on effectiveness and efficacy trials. *Clin Transl Gastroenterol.* 2014 Jan;5(1):e45.

24. Ritenbaugh C, Aickin M, Bradley R, Caspi O, Grimsgaard S, Muscial F. Whole systems research becomes real: new results and next steps. *J Altern Complement Med.* 2010 Jan;16(1):131–137.

25. Carlberg MD, Samuelsson O, Lindholm LH. Atenolol in hypertension: is it a wise choice? *Lancet.* 2004 Nov;264(9446):1684-1689.

26. Klempner MS, et al. Two controlled trials of antibiotic treatment in patients with persistent symptoms and a history of Lyme disease. *N Engl J Med.* 2001;345:85-92.

27. Maron DJ, et al. Initial invasive or conservative strategy for stable coronary disease. *N Engl J Med.* 2020;382:1395–1407. https://www.nejm.org/doi/full/10.1056/NEJMoa1915922?query=RP

28. Bobrow BJ, et al. Chest compression-only CPR by lay rescuers and survival from out-of-hospital cardiac arrest. *JAMA.* 2010;314(13):1447–1454.

29. Mansell NS, Rhon DI, Meyer J. Arthroscopic surgery or physical therapy for patients with femoroacetabular impingement syndrome: a randomized controlled trial with 2 year follow-up. *American Journal of Sports Medicine.* 2018;46(6):1306–1314.

30. Montinari MR, Minelli S, DeCaterina R. The first 3500 years of aspirin history from its roots — a concise summary. *Vascul Pharmacol.* 2019 Feb;113:1–8. doi: 10.1016/j.vph.2018.10.008.

31. Rena G, Pearson ER, Sakamoto K. Molecular mechanism of action of metformin: old or new insights? *Diabetologia* 56(9):1898–1906.

32. Cimolai N. Cyclobenzaprine: a new look at an old pharmacological agent. *Expert Rev Clin Pharmacol.* 2009 May;2(3):255–63. doi: 10.1586/ecp.09.5.

33. Phiel CJ, Klein PS. Molecular targets of lithium action. *Annual Review of Pharmacology and Toxicology*, 2001;41:789–813.

34. Radley DC, Finkelstein SN, Stafford RS. Off-label prescribing among office-based physicians. *Arch Intern Med*. 2006;166(9):1021–1026.

35. Understanding unapproved use of approved drugs "off label." FDA.org. https://www.fda.gov/patients/learn-about-expanded-access-and-other-treatment-options/understanding-unapproved-use-approved-drugs-label. Published February 5, 2018.

36. Hibbert R. What is an immature science? *International Studies in the Philosophy of Science*. 2016;30(1):1–17.

37. Bird, A. Thomas Kuhn. *The Stanford Encyclopedia of Philosophy* (Winter 2018 Edition). Zalta E, ed. https://plato.stanford.edu/archives/win2018/entries/thomas-kuhn/.

38. Eckman P. Traditional Chinese medicine- science or pseudoscience? *Journal of Chinese Medicine*. 2014 Feb;104:41–46.

CHAPTER 4

1. Day R. The origins of the scientific paper: The IMRAD format. *American Medical Writers Association*. 1989;4(2):16–25.

2. Thomas G. *How to do your case study*. Thousand Oaks, CA: Sage Publishing; 2016.

3. Darwin C. *On the Origin of Species by means of natural selection*, 6th ed., 1872.

4. Padilla MJ. The science process skills. Research Matters — to the Science Teacher, No. 9004. Reston, VA: National Association for Research in Science Teaching; 1990. http://www.educ.sfu.ca/narst-site/publications/research/skill.htm.

5. Maciocia G. *The foundations of Chinese medicine*. New York: Churchill Livingstone; 2015.

6. Tal E. "Measurement in science", *The Stanford Encyclopedia of Philosophy* (Fall 2017 Edition), Zalta E, ed. https://plato.stanford.edu/archives/fall2017/entries/measurement-science/.

7. Sabo B, Joffres M, Williams T. How to deal with medically unknown symptoms. *West J Med.* 2000 Feb;172(2):128–130.

8. Birch S. *Japanese acupuncture.* Boston: Paradigm Press; 1998.

9. Papadakis MA, PcPhee S, Rabow MW. *Current medical diagnosis and treatment.* New York: McGraw-Hill Education; 2021.

10. Cheng X. *Chinese acupuncture and moxibustion,* 4th ed. Beijing: Foreign Languages Press; 2019.

11. Furth C, Zeitlin J, Hsiung P. *Thinking with cases: Specialist knowledge in Chinese cultural history.* Honolulu: University of Hawaii Press; 2007.

12. Hanuskins D, Rogers M. *Perspectives: Research & tips to support science education.* NTSA press, 2013.

13. Mannuch AJ, Tusrutani BT, Verkhoglyadova OP, Meng X. On scientific inference in geophysics and the use of numerical simulations for scientific investigations. *Space Science.* 2015;2(8);359–367.

14. Rosenblueth A, Wiener N. The role of models in science. *Philosophy of science.* 1945 Oct;12(4);316–321.

15. Magiorkinis E, Beloukas A, Diamantis A. Scurvy: past, present, and future. *European Journal of Internal Medicine.* 2011;22:147–152.

16. Vallance P, Smart TG. The future of pharmacology. *Br J Pharmacol.* 2006 Jan;147(Suppl 1):S304–S307.

17. Salmon W. *Foundations of scientific inference.* Pittsburgh: University of Pittsburgh Press; 2017.

18. The Discovery of the Double Helix, 1951-1953. NIH: U.S. National Library of Medicine. https://profiles.nlm.nih.gov/spotlight/sc/feature/doublehelix. Accessed July 19, 2020.

19. Gillet J, Varma S, Gottesman M. The clinical relevance of cancer cell lines. *Journal of the National Cancer Institute.* 2013 April;105(7):452–458.

20. Ericcson AC, Crim MJ, Franklin CL. A brief history of animal modeling. *Missouri Medicine.* 2013 May–Jun;110(3):201–205.

21. Bart an der Worp H, Howells DW, Sena ES, Porritt MJ, Rewell S, O'Collins V, Macleod MR. Can animal models of disease reliably

inform human studies? *PLoS Med.* 2010 Mar;7(3):e1000245. doi: 10.1371/journal.pmed.1000245

22. Topol EJ. Failing the public health—rofecoxib, Merck, and the FDA. *N Engl J Med.* 2004;351:1707–1709.

23. Unschuld P. *Nan-Ching: The classic of difficult issues.* Berkeley, CA: University of California Press; 1986.

24. Deadman P. *A manual of acupuncture.* Sussex, England: Journal of Chinese Medicine Publications; 1998.

25. Backman CL, Harris SR. Case studies, single-subject research, and N of 1 randomized trials: Comparisons and contrasts. *Am J Phys Med Rehabil.* 1999 Mar–Apr;78(2):170–176.

26. Chen S, Pollino A. Good practice in Bayesian network modelling. *Environmental modelling and software.* 2012;37:134–135.

27. Bzdok D, Engemann D, Grisel O, Varoquaux G, Thirion B. Prediction and inference diverge in biomedicine: simulations and real-world data. 2018. Hal-01848319.

28. Horvitz E. *From data to knowledge to action: enabling evidence-based healthcare.* A white paper prepared for the Computing Community Consortium committee of the Computing Research Association. https://cra.org/ccc/wp-content/uploads/sites/2/2015/05/From_Data_to_Knowledge_to_Action.pdf.

29. Higginson IJ, Costantini M. Accuracy of prognosis estimates by four palliative care teams: A prospective cohort study. *BMC Palliative Care.* 2002;1(1):1. doi: 10.1186/1472-684X-1-1

30. Burns TW, O'Connor J, Stocklmayer SM. Science communication: a contemporary definition. *Public Understanding of Science* 2003:12;183–202.

31. Odlyzko A. The rapid evolution of scholarly communication. *Learned Publishing,* 2002;15:7–19.

32. Carl Sagan: 'Science is a Way of Thinking.' Science Friday.org. https://www.sciencefriday.com/segments/carl-sagan-science-is-a-way-of-thinking/. Published December 27, 2013.

CHAPTER 5

1. Mann D, Gaylor S, Norton S. Moving toward integrative care: rationales, models, and steps for conventional-care providers. *Complementary Health Practice Review.* 2004 Oct; 9(3):155–172.

2. Rakel D. *Integrative medicine.* 4th ed. New York: Elsevier; 2017.

3. Cowan VS, Cyr V. Complementary and alternative medicine in US medical schools. *Adv Med Educ Pract.* 2015;6:113–117.

4. Zwarenstein M, Goldman J, Reeves S. Interprofessional collaboration: effects of practice-based interventions on professional practice and healthcare outcomes (Review). *The Cochrane database of systematic reviews,* 2009 Issue 3. Art No, CD000072. doi: 10.1002/14651858.CD000072.pub2

5. Ward-Cook K, Reddy B, Mist S. A snapshot of the AOM profession in America: demographics, practice settings, and income. *MJAOM.* 2017;4(4):13–19.

6. Sprague M, Chang J. Integrative approach focusing on acupuncture in the treatment of Complex Regional Pain Syndrome. *Journal of Alternative and Complementary Medicine,* 2011;17(1):67–70.

7. Lihong L. *Classical Chinese medicine.* Hong Kong: Chinese University Press; 2019.

8. Kielczynska BB, Kliger B, Specchio E. Integrating acupuncture in an inpatient setting. *Qualitative Health Research.* 2014;24(1):1242–1252.

9. McCann, JS, ed. *Physician's drug handbook.* 11th ed. New York: Lippincott Williams & Wilkins; 2005.

10. Steel K, Gertman PM, Crescenzi C, Anderson J. Iatrogenic illness on a general medical service at a university hospital. *Qual Saf Health Care.* 2005;13:76–81.

11. Makary MA, Daniel M. Medical error — the third leading cause of death in the US. *BMJ.* 2016 May 3;353:i2139. doi: 10.1136/bmj.i2139

12. Millburn, MP. *The future of healing: exploring the parallels of eastern and western medicine.* Freedom, CA: The Crossing Press; 2001.

13. Juni PJ, et al. Risk of cardiovascular events and rofecoxib: cumulative meta-analysis. *Lancet.* 2004 Dec;364(9450):1995–1996.

14. Birch S. *Japanese acupuncture*. Brookline, MA: Paradigm Press; 1998.

15. Institute of Medicine. *Crossing the Quality Chasm: A New Health System for the 21st Century*. Washington DC: National Academy Press; 2001.

16. Wang SM. An integrative approach for treating postherpetic neuralgia — a case report. *Pain Practice*. 2007 July;7(3):274–278.

17. Tseng YJ, Hung YC, Hu WL. Acupuncture helps regain postoperative consciousness in patients with traumatic brain injury: a case study. *Journal of Alternative and complementary medicine*. 2013;19(5):474–477.

18. MacDonald M. Case study: improving sperm parameters with acupuncture and Chinese herbal medicine. *Australian Journal of Acupuncture and Chinese Medicine*. 2016;10(2):32–36.

19. Lincoln YS, Guba EG. Judging the quality of case study reports. *International Journal of Qualitative Studies in Education*. 1990;3(1):53–59.

CHAPTER 6

1. Wardle J, Roseen E. Integrative medicine case reports: a clinician's guide to publication. *Advances in Integrative Medicine*. 2014 Dec;1(3):144–147.

2. Lang TA. *How to write, publish, and present in the health sciences: a guide for clinicians and laboratory researchers*. Philadelphia: American College of Physicians; 2010.

3. McCann H. Stiff person syndrome. *Journal of Chinese Medicine*. 2008 Feb;86:12–17.

4. Haysom-McDowell AJ, Loyeung BY. Pain with an unexpected outcome for restless leg syndrome. *Australian Journal of Acupuncture and Chinese Medicine*. 2017;11(1):26–30.

5. Schulman D. The unexpected outcomes of acupuncture: case reports in support of refocusing research designs. *Journal of Alternative and Complementary Medicine*. Nov 2004;10(5):785–789.

6. White A. A cumulative review of the range and incidence of significant adverse events associated with acupuncture. *Acupuncture in Medicine*. 2004;22(3):122–133.

7. Institute of Medicine. *Crossing the Quality Chasm: A New Health System for the 21st Century*. Washington DC: National Academy Press; 2001.

8. Epstein RM, Street RL. The values and value of patient-centered care. *The Annals of Family Medicine*. 2011 March;9(2)100–103.

9. Weinstein MC et al. Recommendations of the panel on cost-effectiveness in health and medicine. *JAMA*. 1996 Oct;276(15):1253–1258.

10. Hasnain-Wynia R, Beal AC. The path to equitable health care. *Health Serv Res*. 2012 Aug;47(4):1411–1417.

11. Williams JS, Walker RJ, Egede LE. Achieving equity in an evolving healthcare system: opportunities and challenges. *Am J Med Sci*. 2016 Jan;351(1):33–43.

12. Chao MT, Tippens KM, Connelly E. Utilization of group-based, community acupuncture clinics: a comparative study with nationally representative sample of acupuncture users. *J Altern Complement Med*. 2012 Jun;18(6):561–566.

CHAPTER 7

1. Moore T, Cunningham S ed. *Clinical skills for nursing practice*. London: Taylor and Francis; 2016.

2. Wong-Baker Faces Foundation. www.wongbakerfaces.org.

3. Maciocia G. *The foundations of Chinese medicine*. New York: Churchill Livingstone; 2015.

4. Balke B. A simple field test for the assessment of physical fitness. *Rep Civ Aeromed Res Inst US*. 1963;53:1–8.

5. Bickley LS. *Bates' guide to physical examination and history taking*, 12th ed. Philadelphia: Lippincott Williams & Wilkins; 2016.

6. Zhen LS. *Pulse diagnosis*. Brookline MA: Paradigm Press; 1993.

7. Hammer L. *Chinese pulse diagnosis: a contemporary approach*. Seattle: Eastland Press; 2001.

8. Wang J, Robertson J. *Applied channel theory in Chinese medicine: Wang Ju-Yi's lectures on channel therapeutics*. Seattle: Eastland Press; 2008.

9. Matsumoto K. *Kiiko Matsumoto's clinical strategies*. Natick, MA: Kiiko Matsumoto International; 2002.

10. Weldring T, Smith SM. Article Commentary: Patient-Reported Outcomes (PROs) and patient-reported outcome measures (PROMs). *Health Services Insights*, 2013 Aug. vol.6, https://doi.org/10.4137%2FHSI.S11093

11. Bauer MD. *Making acupuncture pay: real-world advice for successful private practice.* Indianapolis: Dog Ear Publishing; 2011.

CHAPTER 8

1. Patient Consent and Confidentiality. BMJ Author Hub. https://authors.bmj.com/policies/patient-consent-and-confidentiality/.

2. Matsumoto K. *Kiiko Matsumoto's clinical strategies.* Natick, MA: Kiiko Matsumoto International; 2002.

3. MacPherson H, Altman DG, Hammerschlag R, Li YP, Wu TX, White A, Moher D. Revised standards for reporting interventions in clinical trials of acupuncture (STRICTA): extending the CONSORT statement. *Medical Acupuncture.* 2010 Sep;22(3):167–180.

4. Deadman P, Al-Khafaji M, Baker K. *A Manual of Acupuncture.* Sussex, England: Journal of Chinese Medicine Publications; 1998.

5. Riley DS et al. CARE guideline for case reports: explanation and elaboration document. *Journal of Clinical Epidemiology.* 2017;89:218–235.

CHAPTER 9

1. Packer CD, Berger GN, Mookherjee S. *Writing case reports: a practical guide from conception through publication.* Switzerland: Springer International Publishing; 2017.

2. Hofmann AH. *Scientific writing and communication: papers, proposals, and presentations.* New York: Oxford University Press, 2010.

3. Wiseman N. Ye F. *A practical dictionary of Chinese medicine.* Brookline, MA: Paradigm Publications, 2014.

4. Greene BN, Johnson CD. How to write a case report for publication. *Journal of Chiropractic Medicine.* 2006;5(2):72–82.

5. Dehen R. Regression of ductal carcinoma in situ after treatment with acupuncture. *The Journal of Alternative and Complementary Medicine.* 2013;19(11):911–915.

CHAPTER 10

1. Packer CD, Berger GN, Mookherjee S. *Writing case reports: a practical guide from conception through publication.* Cham, Switzerland: Springer International Publishing; 2017.

2. Skelton JR, Edwards SJ. The function of the discussion section in academic medical writing. *BMJ.* 2000 May 6;320(7244):1269–1270.

3. Docherty M. The case for structuring the discussion of scientific papers. *BMJ.* 1999 May 8;318(7193):1224–1225.

4. Dehen R. Regression of ductal carcinoma in situ after treatment with acupuncture. *The Journal of Alternative and Complementary Medicine.* 2013;19(11):911–915.

5. Jefferson T. Restructuring the discussion of scientific papers. *BMJ.* 1999 Aug 28;319(7209):580.

6. Haake M et al. German acupuncture trials (GERAC) for chronic low back pain. *Arch Intern Med.* 2007;167(17):1892–1898.

7. Hayson-McDowell AJ, Loyeung, B, Walsh S. Case study: gua sha for shoulder pain with an unexpected outcome for restless leg syndrome. *Australian Journal of Acupuncture and Chinese Medicine.* 2017;11(1):26–30.

8. Janz S. Plantar fasciitis, another approach — using acupuncture and looking beyond the lower limb with a brief review of conventional care: a case series. *Australian Journal of Acupuncture and Chinese Medicine.* 2010;5(2):30–36.

9. Institute of Medicine. *Crossing the Quality Chasm: A New Health System for the 21st Century.* Washington DC: National Academy Press; 2001.

CHAPTER 11

1. Lang TA. *How to write, publish, and present in the health sciences: a guide for clinicians and laboratory researchers.* Philadelphia: American College of Physicians; 2010.

2. Sun, Z. Tips for writing a case report for the novice author. *J Med Radiat Sci.* 2013 Sep;60(3):108–113.

3. Vinjamury SP. Writing a case report. *Amer Acupunct.* 2008;43:18–19.

4. Chiu E. Psoriatic arthritis managed with multiple styles of acupuncture: a case report. *Meridians Journal of Oriental Medicine.* 2016;3(4):17–22.

5. Anastasi JK, Capili B. The treatment of constipation-predominant Irritable Bowel Syndrome with acupuncture and moxibustion: a case report. *Journal of Chinese Medicine.* 2012 June;99:68–71.

6. Wardle J, Roseen E. Integrative medicine case reports: A clinician's guide to publication. *Advances in Integrative Medicine.* 2014 Dec;1(3):144–147.

CHAPTER 12

1. Gagnier J et al. The CARE Guidelines: Consensus-based Clinical Case Reporting Guideline Development. *Global Advances in Health and Medicine.* 2013 Sept;2(5):1–6.

2. Akers KG. New journals for publishing medical case reports. *J Med Lib. Assoc.* 2016;104(2):146–9.

3. Wardle J, Roseen E. Integrative medicine case reports: A clinician's guide to publication. *Advances in Integrative Medicine.* 2014 Dec;1(3):144–147.

4. *Uniform requirements for manuscripts submitted to biomedical journals.* Philadelphia (PA): International Committee of Medical Journal Editors (ICMJE) Secretariat; updated 2006 Feb. http://www.icmje.org/.

5. Greene BN, Johnson CD. How to write a case report for publication. *Journal of Chiropractic Medicine.* 2006;5(2):72–82.

6. Liesgang TJ, Bartley GB. Toward transparency of financial disclosure. *Opthalmology.* 2014 Nov;121(11):2077–2078.

7. Kelly J, Sadeghieh T, Adli K. Peer review in scientific publications: benefits, critiques, and a survival guide. *EJIFCC.* 2014 Oct;25(3):227–243.

8. Jefferson T, Alderson P, Wager E. Effects of editorial peer review: a systematic review. *JAMA.* 2002;287(21):2784-2786.

9. Smith R. Peer review: a flawed process at the heart of science and Journals. *JR Soc Med.* 2006 Apr;99(4):178–182.

10. Lang TA. *How to write, publish, and present in the health sciences: A guide for clinicians and laboratory researchers*. Philadelphia: American College of Physicians; 2010.

APPENDIX A

1. Birch S, Felt R. *Understanding acupuncture*. New York: Churchill Livingstone; 1999.
2. Birch S, Ida J. *Japanese acupuncture: A clinical guide*. Brookline, MA: Paradigm Publications; 1998.
3. Shudo D. *Japanese classical acupuncture: Introduction to meridian therapy*. Seattle: Eastland Press; 1990.
4. Kodo F. *Meridian Therapy: A hands-on text on traditional Japanese Hari based on pulse diagnosis*. Japan: Toyohari Medical Association; 1991.
5. Kobayashi S. *Acupuncture core therapy: shakujyu chiryo*. Brookline, MA: Paradigm Publications; 2008.
6. Matsumoto K, Euler D. *Clinical strategies of Kiiko Matsumoto: Vol 1*. Natick, MA: Kiiko Matsumoto International; 1999.
7. Birch S. *Shonishin: Japanese pediatric acupuncture*. New York: Georg Thieme Verlag; 2011.
8. Tan R, Acupuncture 1, 2, 3. Self-Published, 2007.
9. Young, WC. *Thoughts of Wei-Chieh Young: Developments and Expansions of the Principles and Theories of Master Tung's Extraordinary Points*. Rowland Heights, CA: American Chinese Medical Culture Center; 2014.
10. Wang Y. *Micro-acupuncture in practice*. St. Louis, MO: Churchill Livingston Elsevier; 2009.
11. Yoo TW. *Koryo Hand Therapy, Vol 1*, 2nd ed. Korea: Eum Yang Mek Jin Publishing Co.; 2001.
12. Park M, Kim S. A Modern Clinical Approach of the Traditional Korean Saam Acupuncture. *Evidence-Based Complementary and Alternative Medicine* 2015. https://doi.org/10.1155/2015/703439
13. Worsley JB. *Description of Worsley Five-Element Acupuncture*. Personal communication, October, 2020.

14. Reaves W. *The acupuncture handbook of sports injuries and pain: A four step approach to treatment.* Boulder, CO: Hidden Needle Press; 2011.

15. Callison M. *Sports medicine acupuncture: An integrated approach combining sports medicine and traditional Chinese medicine.* San Diego: Acusport Education; 2019.

16. Dommerholt J. Dry Needling — peripheral and central considerations. *J Man Manip Ther.* 2011 Nov;19(4):223–227.

APPENDIX C

1. Protection of Research Participants. International Committee of Medical Journal Editors. http://www.icmje.org/recommendations/browse/roles-and-responsibilities/protection-of-research-participants.html.

2. Battershill P. Should informed consent be required to publish a case report? *Canadian Journal of Hospital Pharmacy.* 2007 Nov;60(5):335–336.

3. Barbour V. on behalf of COPE Council. Journals' Best Practices for Ensuring Consent for Publishing Medical Case Reports: guidance from COPE December 2016. https://publicationethics.org.

APPENDIX E

1. Chiu E. Using a rubric to evaluate quality in case study writing. *Meridians Journal of Acupuncture and Oriental Medicine.* 2017;4(4):31–40.

ACKNOWLEDGMENTS

I would like to thank all of the mentors and teachers I have had the fortune to learn from over the years, from the New England School of Acupuncture, at the Oregon College of Oriental Medicine, during my time spent in Taiwan, and in the professional community. Thank you for generously sharing your knowledge and wisdom of this medicine. I am grateful also to my students for inspiring me with your questions and with your commitment, to caring for your patients but also to writing to share your experience for the benefit of the field.

I would like to thank my editor, D. Olson Pook, for your expertise and dedication in helping me with the various aspects of bringing this book together, from structural organization to the smallest details. Thank you to Lynn Eder, for your careful work with the final draft and with proofreading. Also, thanks to Sunjae Lee for the stunning artwork on the cover, and Cecile Kaufman for brilliant book design.

Thank you to my colleagues and friends who helped me to review chapters and provide feedback and encouragement as I wrote: Selma Jones, Bob Quinn, Anthony Chen, Robert Kaneko, Joe Coletto, Lee Hullender-Rubin, and Ken Glowacki. At OCOM, thanks to Beth Burch, Zhaoxue Lu, Henry McCann, Candise Branum, and the administrators and faculty. At Vancouver Acupuncture, thanks to Andy Kelch and Jana Peterson.

Special thanks to the many leaders in the field who consulted in verifying the accuracy of the descriptions of the various styles of acupuncture: Wei-Chieh Young and Alice Young (www.drweichiehyoung.com), Delphine Armand and Paul Wang (www.siyuanbalance.com), Kiiko Matsumoto and Monika Kobylecka (www.kiikomatsumoto.com), Judy Worsley and David Berkshire (www.worsleyinstitute.com), Whitfield Reaves, Peter Eckman, Toby Daly, Zoe Brenner, Diane Iuliano, and Ken Glowacki.

And lastly, my family. To my parents, for starting me on this journey and for your support along the way, my brothers David and Stephen for your encouragement and advice, and finally to my wife Yuchin and my son Justin for your patience, support, and love. Thank you.

ABOUT THE AUTHOR

Edward Chiu, LAc, DAOM has been teaching acupuncture case report writing for over ten years as a faculty member at the Oregon College of Oriental Medicine and has also served on the faculty at the National University of Natural Medicine and the American College of Traditional Chinese Medicine. He earned his undergraduate degree in biology at Harvard University and worked several years as an educator at the Museum of Science in Boston developing programs and exhibits encouraging visitors to engage in scientific skills. After earning his Master's in Acupuncture from the New England School of Acupuncture and his Doctorate in Acupuncture and Oriental Medicine from OCOM, he spent time in apprenticeships studying acupuncture in Taiwan and has also been involved in acupuncture research studies in the US at both Massachusetts General Hospital and the University of Arizona. He has over two decades of private practice and represents the third generation in his family to follow in the tradition of practicing Chinese medicine.

For resources relevant to writing case reports in acupuncture,
please visit the website

www.acupuncturecasereports.org

9 781735 958309